THE BOOK OF SEARCH

BY FIREFIGHTERS, FOR FIREFIGHTERS

The PDF is free to all. Share it with everyone.
The proceeds from the sale of all physical copies
goes to:

The Firefighter Rescue Survey

You can also directly support the project by following the QR code below to become a member of the FFRS Patreon Community.

All articles, artwork and photos in this book belong to their respective owners. Firehouse Vigilance is the author of the Book of Search but has no rights to the material other than the donated use for publication in this document.

Firehouse Vigilance 2024

Foreword by David Rhodes
Edited by Corley Moore
Authored by Firehouse Vigilance

Some photographs were submitted by the authors for their articles. If no credit is mentioned, the shots belong to the author of the article.

For those articles that did not have photos. The following artist were gracious enough to donate their galleries.

Chris Smead
https://www.csmeadphotography.com

Jillian Stewart
https://www.ishootfire.com

Samantha Worthy
https://cameraworthyphotography.smugmug.com/

FOREWORD BY DAVID RHODES 8

SECTION 1: MINDSET 11

    WHY WE EXIST 13

    THE VIGILANT MINDSET 17

    CRUCIBLE MOMENTS 21

    THE IMPACT OF TIME 25

    THE SEARCH: A CHIEF'S PERSPECTIVE 29

    SEARCH CULTURE 33

    SEARCH AND COMMON SENSE 37

    THE WHY OF ENTERING WINDOWS 41

    PRIMARY SEARCH VS EMERGENCY RESCUE MODE 45

    RECORDING OUR RESCUES TO DRIVE OUR MISSION 49

SECTION 2: PREPARATION 53

    TRAINING FOR SEARCH 55

    GROUND LADDERS FOR RESCUE 61

    TRAINING AGAINST THE CLOCK 65

    THERMAL IMAGING TIPS FOR SUCCESS 69

    SEARCH: POSITIONS AND POISE 77

    EFFECTIVE ARTIFICIAL STRESS 81

    MASKING UP 87

    LITHIUM ION & SEARCH 93

    THE PSYCHOLOGY AND PHYSIOLOGY OF SEARCH 97

    THE DIRTY SIDE OF THE PRIMARY SEARCH 101

SECTION 3: DECISIONS 105

    KEEP SAYING YES TO V.E.S. 107

    RULES OF ENGAGEMENT FOR SEARCH 113

    SEARCHING WITHOUT A HOSELINE 117

    SEARCHABLE SPACE: THE SIZE-UP 127

    THE FIRST-IN INTERVIEW 131

    ASKING THE RIGHT QUESTIONS 135

    COMMAND FOR THE RESCUE 139

    COMMUNICATION CONSIDERATIONS 147

    LARGE AREA SEARCHES 155

    THE IMPACT OF ERA ON SEARCH 161

    HIGH-RISE SEARCH & RESCUE 169

    OPTIMIZING VENTILATION FOR SEARCH 173

SECTION 4: ACTION 179

    CIVILIAN DRAGS 181

    CONSIDERATIONS FOR VICTIM REMOVAL 187

    WINDOW BASED RESCUE 191

    SEARCHING IN HOARDER OCCUPANCIES 201

    WINDOW CONVERSIONS 205

    CAN CONFIDENCE: 211

    ORIENTED VS SPLIT SEARCH 217

    ORIENTED SEARCH 225

    SPLIT SEARCH 231

    LIFE, FIRE, LAYOUT (AND CALL OUT!) 235

    THE SUBURBAN ENGINE SEARCH 241

    TOWER LADDERS: RESCUES & REMOVALS 245

    SEARCHING OFF THE INITIAL ATTACK LINE 251

    SECONDARY SEARCH 255

AUTHOR GLOSSARY 259

THE ARTISTS 282

    HOW THE BOOK OF SEARCH CAME TO BE 285

FOREWORD
DAVID RHODES

During my career, I witnessed several periods of change regarding the priority of the primary search. Starting in a small suburban department with low staffing levels, our priority was a quick knockdown of the fire using a preconnected line and booster tank water, then dropping the line and getting right to the primary search. This was usually done alone or with one other firefighter during the initial 5-10 minutes.

Leaving that environment and going to the urban environment in the early 1990s, the staffing levels and response times were such that multiple tasks could be accomplished simultaneously. Unfortunately, the practice at that time was to get the fire totally under control, then search in several battalions and shifts. In the late '90s, we started to shift back to the first truck being assigned primary search. There was more training on VES and, as word spread of more and more rescues being made, more command staff members embraced the tactic. Predetermined tactical assignments even made it into the operations manual, with an emphasis on the primary search being the priority.

With staffing levels and resources being so different across the many jurisdictions, each local department must develop their operational procedures based on the resources they typically deploy. At the heart of those procedures must remain an understanding that life safety and the rescue of anyone trapped are our top priorities. If policy and training are developed with that in mind, then firefighters should arrive on scene with the ability to quickly size-up a situation and take the action that best fits the situation and resources. In some cases, that will mean fast water to knock down the fire so that you can search. In other situations, it could mean that you enter a window and close the door to effect a search of a targeted area without a hoseline or an additional crew on scene to back you up with one. In a perfect world where the resources arriving are sufficient, we deploy a hoseline for fire attack and interior exposure attention while separate crews enter from multiple locations to search.

The fire service's mission for more than 200 years has been to save lives and property. With the expansion of insurance and plentiful building supplies, the protection of most property does not carry the same risk/benefit as saving lives. We know the structures will most likely be torn down and replaced, depending on the severity of the damage. Saving life, on the other hand, is worth the level of risk associated with the action. If we approach any incident in which people occupy the space with the mindset that we are there to accomplish search and every other action and activity we engage in is to support the search effort, then we can get in, do what we need to do, and get out as quickly as possible.

We are learning now with actual data that our initial actions in the first 4-8 minutes can pay huge dividends when it comes to the survival of trapped victims. This information has reignited the search conversation and is forcing organizations to adjust tactics and reexamine policy and training. We are also fortunate to be here in a time of critical research on fire dynamics, the flow of toxic gases throughout a structure, the effects of ventilation, and water application and what it does to the environment. We have learned that a closed door may be just as effective as a hoseline in buying us time to search and limit toxic gas exposure to a victim. While all this new information is giving us a better understanding of the interior environment, we are also confirming what training manuals have stated for a century: Time is against the victim.

This book is a collection of some of the brightest minds in our business who are established as true change agents. Their voices are important as we transition from a 20-year culture of risk averse leadership and policy that gripped our industry and forgot time was against the victim. It is time to focus on our primary mission of saving lives and educate the fire service leaders that making search a priority is NOT a choice between operating safely and recklessly. Understanding building construction, fire dynamics, human behavior, ventilation, and water application and being proficient in every type of search tactic are the highest levels of risk management. It requires advanced and continuing training and a commitment to the people your organization protects. It requires following the mission of saving lives. It requires competence and character. It in no way requires you to operate recklessly. We can never remove all risks, but through education and training we can build the understanding that it takes for sound decision making.

As you digest the material in this collection, always keep in mind that firefighting and tactics are local, based on your staffing and resources. The science of fire dynamics, ventilation, water application, and victim exposure to toxic gases, however, is the same everywhere. Prioritizing the life safety of those we are sworn to protect should guide organizational policy, procedure, and culture. Staying current on research, contributing data to the Firefighter Rescue Survey, and making sure every firefighter is equipped with understanding and the why of every tool, tactic, and strategy will lead to the saving of lives in our communities.

David Rhodes
Editor in Chief
Fire Engineering

UNDERSTANDING THIS BOOK

All the articles and drills that you will read in this book have been donated by their respective authors. While editing this book I have worked hard to vet every article and make sure that all the information is validated. However, it must be understood that some of the articles are strictly the opinion of the author. Not everyone will agree with everything in this book, and that is ok. Discourse and healthy disagreement are what will drive the fire service forward to improvement and best practice. Take the bits and pieces that you like and apply them to your department. Analyze the parts that you disagree with and let them reinforce your beliefs, or maybe challenge and change what you currently believe.

The book is laid out in four sections, each section covers a specific aspect of search.

Section 1: Mindset
Section 2: Preparation
Section 3: Decisions
Section 4: Action

This book is not designed to be read from front cover to back cover. It is laid out so that you can flip to whatever aspect of search you want to study and read an article on a topic related to that aspect. There is an author glossary in the back so you can locate an article by who wrote it. This hardback version does not contain the drills section that you will find in the PDF. (Don't worry a printed drill supplement is coming.)

My hope is that the Book of Search will be an invaluable tool for the fire service, and the best part is that it is 100% free. So, make sure you share it with as many people as you can. It was an honor to work on it, and it was a herculean effort by so many people.

Enjoy the journey. Godspeed.

Corley Moore

SECTION 1: MINDSET

The articles in this section cover the mental aspect of search. The topics are all about your search philosophy and the mindset you have when approaching the training for and the execution of search.

WHY WE EXIST
MIKE LOMBARDO

"The hallmark of a good fire department is an aggressive and thorough search. Most departments mention somewhere in their mission statements the tenet to save life and property. In reality, saving a life is what firefighters are here for."

~ Chief Mike Lombardo

THE FOLLOWING IS AN ACCOUNT OF A FIRE THAT I RESPONDED TO MANY YEARS AGO:

On the evening of January 29, 1998, at approximately 6:30 p.m., a full first-alarm assignment was dispatched to a report of a fire on Townsend Street in Buffalo, New York. The assignment consisted of three engine companies, two truck companies, a rescue company, and a battalion chief.

Truck 11 arrived right behind Battalion 3; the fire was only two blocks from the unit`s quarters. It is a single unit stationed only with the chief; it carries no water and was staffed that evening with five firefighters and an officer. On arrival, the fire was observed venting from two doors and two windows on the number 4 side, from the first-floor rear apartment of this two-story wood-frame dwelling.

With very heavy fire venting from every opening on the number 4 side of the building except one and no engine company yet on location, the prudent decision would have been to await the arrival of an engine and the stretching of a line. However, there were also a frantic mother and father screaming that one of their children was not yet out of the apartment.

Battalion Chief Tom McNaughton also relayed to us that a child was indeed inside the building. He requested that we attempt to enter and search for the child.

There were no openings on the number 3 side of the structure, and windows on the number 2 side were immediately inaccessible by security bars (doors to the apartment were on the number 4 side).

I made the decision to enter the only remaining window into the apartment that was not venting fire. Heavy smoke pushed from the window. Firefighters Tom Jackson and Chuck Sardo and I entered the window into a bathroom. There was a high heat condition in this room. Ahead was a small hallway, where fire was rolling across the ceiling. Jackson crawled through the hallway and into the kitchen. Conditions were worsening rapidly. Fire was heavy in the kitchen.

Outside, Truck 11's driver, Firefighter Tom Schmelzinger, handed a 2nd water can into the bathroom window to me while Firefighters Tom Sullivan and Mike Taube went to the number 2 side of the building to force entry through the security bars on the windows there. (There were also scissor gates on the doors of this apartment house, though they were not a factor in the fire.)

Jackson traveled through the kitchen, with Sardo and me following. I tried to protect them as much as possible with the water can. Then Jackson entered a small bedroom off the kitchen. He searched a set of bunk beds in this room, with negative results. He came to a pile of clothes in front of the bedroom closet. He found a two-year-old boy.

The bedroom window was barred, providing no exit. Jackson rushed the baby out of the room and almost became trapped in the tiny space at the beginning of the hall between the kitchen sink and hallway wall, which measured less than 18 inches. His helmet was dislodged halfway off his head. He handed the baby to Sardo, who handed the child to me, and I passed him outside to firefighters. The child was in cardiac arrest, and the firefighter's performed CPR as they rushed him to a waiting police car.

Meanwhile, I used the 2nd water can protect Jackson and Sardo as they made their way forward to the bathroom. It did not extinguish much fire but slowed its progress. I ascertained from Chief McNaughton that this was the only person reported to be in the structure, and we exited the structure. Engine 3's crew had advanced a line into the building by this time and pushed into the apartment, quickly controlling the fire.

WHAT WOULD YOU DO?

There was tremendous heat in the bathroom, where our team entered. The tub surround had melted into the bathtub, and a medicine cabinet had melted off the wall. Firefighter Jackson received minor burns to his head when his helmet was dislodged in the hallway. These types of conditions normally would indicate that entry should not be made without a handline.

However, with reliable reports such as those given that evening by the child's family, an attempt must be made to enter and search. If a handline had been immediately available, it

SECTION 1: MINDSET

Photo by: Camera Worthy Photography

still may not have guaranteed success; it most likely would have been advanced in through the apartment door, and crews would have had to delay the search while this line was advanced.

About two months after this fire, a man and woman walked into the quarters of Truck 11. With them was their son, Elijah, the boy rescued from the fire. The child had a large burn on his head that was still healing, but otherwise he was in great shape. If his parents were asked about the firefighting risk vs. benefit of the rescue of their child, there is no question what their answer would be. And with the successful rescue of the baby, I am sure that the collective fire service voice agrees.

At the time we entered, Elijah Hall`s life was in the balance, and the duration of that life would be decided within the next few seconds.

But what happens when the child does not survive, or a firefighter does not survive or is seriously injured? It seems, then, that the collective fire service voice is very muddled with armchair quarterbacks saying, "I told you so." The tenets of victim profiling applied to this incident surely would have indicated FFs should not enter and that the space is not survivable.

WHY WE EXIST – M. LOMBARDO

> *Decisions such as the one made on Townsend Street are not made by a computer or in a classroom with time to ponder. They are made in a split second and often without complete information. Elijah Hall's life was saved primarily by the actions of Firefighter Tom Jackson, but also in part by all the members of the team of firefighters who responded that evening. He was saved because Tom–with his training and experience and his team behind him, fully recognizing the risk "went out and did what he had to do." And that's the essence of the fire service.*

THE RULES & THE ODDS

Risk analysis models influence much of the fireground decision making in the fire service today. But at times we are called to go against these models, act against the odds. The results of such actions are sometimes tragic and sometimes successful. Regardless of the outcome, the fire service must remember that we are a human service, and a standard set of rules or guidelines cannot always dictate the actions of the firefighters who serve the public.

We should search every building every time. Remember, a neighbor saying no one is home does not constitute a primary search. Even if we can't enter immediately, we must search as soon as we can. It is unacceptable for the fire investigator, or the board-up crew or God forbid the family to find the body of their loved one.

WHY WE EXIST

Incident Commanders often give an arriving unit the assignment of search. I don't look at search as an assignment but rather a mission to SAVE LIVES, and that is everyone's mission.

Now don't misunderstand me when I say this, firefighters are performing other tasks on the fireground, but they are all directed to the mission to SAVE LIVES!

We perform forcible entry to provide entry to SAVE LIVES, we ventilate the building to improve visibility and give victims hopefully some better air to SAVE LIVES. We ladder the building not to "empty the ladder bay" but to provide points of entry and egress to SAVE LIVES. We stretch and operate hoselines so we can knock down and extinguish fire to SAVE LIVES. Everything we do on the fireground should be directed towards SAVING LIVES.

I hear reference made today to "victim profiling", I reject that concept. I am going to give it my best to get in to search and if I am pushed back by fire I am going to keep trying until I can get in. I am not, as a FF, as a company officer, or as a command officer ever going to simply write off people and say, "no one can survive in there". Victim profiling and playing God is not my job, saving people is.

THE VIGILANT MINDSET
DAVE LEBLANC

Photo by: Csmeadphotography

For years now the catch phrases "Expect Fires" and "Expect Victims" have been used to foster a mindset that we as firefighters should have every time we go out the door to a reported fire. Expect fire was born from the Andy Fredericks quote about the garbage man not getting excited when he sees a pile of garbage.

"A garbage man doesn't get excited when he turns the corner and sees trash because he is expecting it. Likewise, you should be expecting fire on every run." – Lieutenant Andy Fredericks – FDNY Squad 18

Now, no doubt there is a vast difference between seeing a pile of cardboard on the sidewalk and pulling up to fire showing from three windows, and you are bound to feel some level of excitement. But it is a possibility, and keeping your mind focused on that possibility will help with your ability to function. However, maintaining that level of vigilance isn't easy, especially when we respond to far more false alarms than actual fires. It is easy for our level of complacency to increase and for us to let our guard down. The result is never good for the public and rarely good for us.

Complacency allows us to get caught by surprise when we pull up to fire showing. This lack of mental preparation affects our ability to perform a good size up and also our ability to act with purpose. It's been proven that hormonal heart rate increase cause both our cognitive and psychomotor abilities to suffer.

BIAS TO ACTION

The Fire Service is predicated on a Bias to Action that is the tendency to favor action over inaction. Our careers are literally built preparing to act at any given moment in a capacity that is filled with risk for our citizens as well as us. This preparation is partly physical, as we must prepare ourselves to execute the skills necessary to search, vent and extinguish fire. But this preparation is also mental, where we prepare our minds for where our bodies must go. This mental conditioning also enables us to not be "surprised" by what we encounter at any given incident. That lack of surprise allows us to control our emotions and operate effectively.

Our surprise can lead to fear or hesitation. It can lead us to make bad decisions. It can lead to us being ineffective, or more importantly it can lead to our deaths.

ACT LIKE YOU HAVE BEEN THERE BEFORE

When we show up on the fireground, we need to be able to perform like we have trained for. We shouldn't have to consciously think of all the steps required to perform a specific task, they should be part of our muscle memory, allowing our mind to focus on the scene in front of us and constantly evaluate conditions and actions needed.

JD Ducharme was on the Make Due Suburban Firefighter podcast and discussed our risk management techniques. He more appropriately dubbed it risk assessment. We don't manage our risk, although some of our actions will reduce it, instead we assess the risk involved with the next task we need to perform and then we weigh that risk against our abilities and resources to determine if we can accomplish the task.

Much like "Expect Fire", the phrase "Expect Victims" was born from the same thought process. We should respond to building fires with the mindset that there will be victims in the building. The fire started somehow, and the newspapers are filled with stories of firefighters pulling occupants from "vacant" buildings, and citizens surviving unthinkable conditions because firefighters made the decision to act as they were trained and search the building for life.

It is critical as firefighters we don't lose sight of why we are here, and while our safety is important, we swore an oath to protect the lives and property of the citizens we serve. That does not

Photo by: Csmeadphotography

mean we recklessly risk our lives, but we should accept the risk that comes with our job and rely on our training to keep us safe while upholding the oath we took.

"Our jobs are not to be safe but to be effective and do that as safely as possible." ~ Senior Chief Robert Howell, United States Navy SEALS

Being a firefighter in today's fire service can be compared to being a backup quarterback on an NFL football team. You practice during the week, take snaps and execute plays; and then you sit on the bench and wait for your moment. When your moment comes, you are expected to perform that same as the starting quarterback. It is the same team, all the other players are the same, the backup just has to stand in and execute. Much like you, as a firefighter, can train with your shift, and then spend hours at the station or on routine calls, waiting for your moment to execute at a building fire, falling back on your training and experience to perform

SECTION 1: MNDSET

your job. If you are spending that time preparing for your moment— and more importantly if your mindset isn't expecting that you will be going to work when you respond to that reported fire, then you are setting yourself up for failure. More importantly you are setting yourself up to fail to meet the expectations of the very public you swore to protect. If we prepare ourselves physically to meet the challenges of firefighting, we are building in our safety. If we prepare ourselves mentally to expect a fire on every run, and to expect victims at those fires, then we will operate from a position of confidence. Confidence will erase the fears and doubts that come from the unexpected.

CRUCIBLE MOMENTS
JEFF ROTHMEIER

Photo by: Csmeadphotography

A quiet night was a rare occurrence for this fire house…Yet the ominous Tones dropped and a bit more hustle ensued from the whole house than normal. All units dispatched to a structure fire with victims trapped on the roof and in the house... My short time in the front seat and even less time on a truck sent my mind scanning for appropriate ways to lead the troops and ready myself for the moment, one of scarce occurrence in the profession… A rescue. Once on scene everything happened rather fast, and decisions seemed impossible. A significant amount of fire was barking out several windows on the second floor. As our company threw ladders I wondered if the interior stairwell would be a better means to start our search. Eventually information was given of the victim's location in the rear 2nd floor bedroom, our route seemed predetermined as this is where we chose to ladder. We committed to the space and fortunately it was protected by the line, the victim was found rapidly and removed. Yet his condition was not favorable enough to survive.

MOMENT OF TRUTH

A crucible moment for me was described to you in this story. A similar story has happened to other members in the fire service, both young and old. I have seen it play out in various other instances and heard it retold in various forms. The lesson is always the same. We must be all that we can be when the moment of truth encounters us. The humbling reality is such that when our mission confronts us we will realize it in the aftermath. No escape is found in the recesses of our mind, should we know our full potential was not realized. No

Photo by: Csmeadphotography

search is done quick enough post event, it's interesting how easily we are satisfied with our efforts during drill, and yet how critical we can be post-incident. When this moment confronts you, will your remembrance be a justification source for the work you have engaged in preceding the event or a motivating stream of consciousness to ensure you never fall short again. The mark is that of which you set for yourself, the greatest critic.

WHERE WAS YOUR MINDSET FORGED?

Your engagement in this craft is forged in the crucible of fire combat. You won't fully appreciate the weight of your responsibility until you routinely confront the conditions preventing you from reaching your objective. The weight of your mission is more burdensome than you realize. Confront these conditions in earnest. Victims are heavier than you realize, you are not as strong as you think, the fire is hotter than expected, the window is higher than you thought, the home is darker than you would like, and your partner will be of less assistance then you hope. Once you confront the reality of our profession at a scene with a known victim, you will learn the standard for proper conditioning. The only appropriate response for such a paradigm is a humble dedication to competency, the public, and combat conditioning.

WHAT WOULD CHANGE?

If you knew your shift would include a search for an entrapped victim in a structure fire before its conclusion, how would the conduct of your day at the firehouse look. By some miracles let's envision that you were given absolute information which foretold this eventual truth. How has your shift changed for the day? What new habits and behaviors have you implemented? This will be the foundation of our philosophy which is a product of belief, and belief the fuel for purpose. Firefighters with no belief are as effective at lifesaving as a car with no gas. Seek to be relevant on every scene and in each moment it presents, and here you will find waves of combat effectiveness.

YOUR WHY

Your hope of joining our ranks was rooted in this philosophy. The noble intention to save lives, this was the notion that drove you to pursue this craft from the onset. The notion didn't die, it's still the fire service reality, who killed it? I can tell you who didn't kill it, the public's need. They still believe in the fire service's nobility. It's why we drive fire trucks through parades, get thanked in grocery lines, and have sweets delivered to the fire station. The public is grateful for its noble

SECTION 1: MINDSET

servants. What daily intent has guided your actions that seem at odds with the intent to save lives? What intent has robbed you of the physical strength to perform rescues in full gear? What intent has distracted you from conducting effective size ups on building layouts? What intent has robbed your initiative? Your profession is not one which favors the constructs of half measures. It's time we reinvest in a philosophy which will support our intention. The one which leads us to fulfillment. We are fire combatants, our intent is on killing fire, saving lives and protecting property. As such, Search is every firefighter's core responsibility

THE IMPACT OF TIME

BEN SHULTZ

Photo by: Jillian Stewart - iShootFire

The fire service exists to provide the public with the best chance for the best possible outcome. This is the mission.

To uphold this mission, we must acknowledge that there are far more possible outcomes than simply life or death. Yes, the outcome of death is absolute. There are no varying degrees of death, and sometimes, no matter what we do, no matter how well prepared we are and how efficiently we perform, the citizen is not going to make it. However, when the outcome is life, our citizens survive in the gray. There is a significant difference in quality of life for a patient with second degree burns to 10% of their body versus third degree to 20%. Additionally, there is a significant difference in quality of life and brain function for a citizen that has a three-minute anoxic or hypoxic brain injury versus one that lasts six minutes.

We do not show up to fires simply to remove someone from the building. To operate in such a manner, or to utilize this task as a benchmark or gold standard, would be the search and rescue equivalent of "the fire went out." Our training, education, and equipment must support operations that value, not only our role in victim removal, but our role in quality of life. The unprotected citizens are up against multiple factors that determine their outcome. Of these various factors, the only one that the fire service can impact, both positively and negatively, from tones to treatment is time.

5 FACTORS OF PATIENT OUTCOMES

The following factors conspire to determine victim outcomes in a structure fire. The fire department has a varying degree of influence on each of these, from none to significant control.

#1: CO-MORBIDITIES

These are the pre-existing conditions of our patients including, but not limited to, their age, weight, previous illness, and previous injury. The fire department has no say in these factors.

#2: TREATMENT

Includes treatment administered via EMS on scene, in the hospital, and long-term rehabilitation. If the fire department is responsible for EMS, then they may have significant impact on the factor of treatment. However, the ability to begin treatment does not begin until the patient has been removed from the IDLH.

#3: PATIENT LOCATION

Location refers to the patient's proximity to the heat and/ or gasses; their elevation within the space, i.e. on the floor, on a bed, on a bunk bed; and their degree of isolation from the fire and its byproducts, i.e. behind a closed door or in an open room. The fire department cannot begin to impact these factors until we either locate the patient or begin coordinated attack and ventilation.

#4: CONDITIONS

Refers to the temperatures, the smoke density, and contents of the smoke and fire gases. Once again, the fire department cannot begin to impact conditions until attack and ventilation operations begin.

#5: TIME

Includes duration of exposure to the conditions and time to receive appropriate medical interventions. This is the one and only factor that the fire department can impact prior to arriving on scene and throughout their interaction with the patient.

We do not get to pick our patients. We cannot begin to impact treatment, patient location, or conditions until we are on scene and operations are underway. However, our ability to impact time

Photo by: Jillian Stewart - iShootFire

begins the second we are notified of the call and continues all the way through treatment. With the unrelenting, linear influence of time on patient outcomes (death vs. quality of life), it is imperative that fire departments scrutinize their operations second by second.

WHAT THE DATA SAYS

Unfortunately, the fire service is in its infancy in terms of data collection and analysis when it comes to structure fire patient outcomes. The two most prevalent sources of data, currently, are UL-FSRI and Firefighter Rescue Survey (FRS). These entities provide very different data from very different sources, but what they both show is that the unprotected public has a very small, rapidly diminishing window for an outcome other than death.

SECTION 1: MINDSET

Photo by: Jillian Stewart - iShootFire

The UL-FSRI Study of the Impact of Fire Attack Utilizing Interior and Exterior Streams on Firefighter Safety and Occupant Survival showed us that in three different structure fire profiles (2 rooms of fire/ 2 vents; 1 room of fire/ 1 vent; 1 room of fire/ no vent) the same two victim locations reached doses of either heat or toxic gasses that would be lethal to 50% of the population in a time period between 3 minutes 6 seconds and 8 minutes 10 seconds, from the time the fire started. In and of themselves, these times are miniscule windows of survival. And they only get smaller when we begin to account for time for dispatching, turnout time, and response time. Unfortunately, this study alone is often not enough to convince everyone that measuring our performance down to each second is necessary. Firefighters are often quick to point out that these are "lab values" in controlled settings and may not represent the real world in which we operate.

To counter those concerns, thankfully, we can look at the work of the Firefighter Rescue Survey. Through the collection of its first 3000 rescue surveys, the Survey has shown that the longer it took to locate a patient, the lower the rate of survival has been. More specifically, patients located within the first 2 minutes of arrival of the first due unit had an 81% survival rate. This number drops substantially to 69% when locating the victim took between 2-4 minutes, and then continues to drop 5% for every additional 2 minutes it took to locate patients.

While the Survey lacks the controls and instrumentation of a UL-FSRI study, it is providing numbers from real patients removed from real fires.

The takeaway is this— regardless of the source of data – the "lab values" of UL-FSRI or the "street data" from Firefighter Rescue Survey— they both clearly display two key concepts: a correlation between time and patient survival rates; and the window of survival is small and closing quickly.

THE 4T's OF TIME LOSS

Efficient operations are often incorrectly referred to as "buying back time". We must acknowledge that time cannot be recovered. Time lives in an unrelenting, linear space. We can only hope to achieve the goal of minimizing the time required to complete a task. Embracing the linear nature of time should lead us to evaluate our operations in linear fashion, from tones to treatment.

The four distinct phases of our operations, regarding time (in chronological order), are as follows:

#1: TURNOUT TIME
From the second we are alerted to the incident until we are driving out of the station. Turnout time is impacted by how quickly we get to the apparatus, how quickly we get dressed, and how quickly the bay doors open.

#2: TRAVEL TIME
From driving out of the station, until we have positioned our apparatus on scene. Travel time is impacted by our path of travel to the scene, our water supply decisions, and our apparatus positioning.

#3: TAILBOARD TIME

From exiting the apparatus until we are walking away from the apparatus with the tools we need for the assignment we are given. Tailboard time is impacted by apparatus design, equipment choices, equipment storage, and economy of motion.

#4: TASK TIME

From walking away from the apparatus until we complete our given assignment. Task time is impacted by knowledge, training, equipment, proficiency, and staffing.

Photo by: Jillian Stewart - iShootFire

EMBRACE THE STOPWATCH

When evaluating our performance, it is imperative that we examine all four phases above, as multiple operational inefficiencies lie within each one. It can often seem like a daunting task to attempt to be one minute faster at an operation. However, asking a crew to save 15 seconds in each of the four phases often seems much more realistic and palatable. Continuous small victories along the linear path of time lead to big wins in overall operations.

To effectively evaluate the times above, we must bring the stopwatch back into our training. It is quite simply the best metric we possess to determine whether our changes are improving our operational speed or not. Additionally, putting a clock to our operations allows us to differentiate between those that are just effective versus those that are effective *and* efficient. Effective operations only focus on the result. The door gets forced. The primary gets completed. The roof gets laddered— but 'effective' fails to account for the amount of time it took to complete the task. Efficient operations achieve the desired result and do so with the least amount of time required.

Keep in mind, the stopwatch does not have to be punitive. Additionally, the stopwatch should not be used when learning a new skill. But as we progress to the "walk" and "run" phases of learning, the stopwatch should be employed in your operational evaluations.

SPEED VS. RUSHING

There is a massive difference between getting faster through economy of motion and rushing. We are not attempting to get faster or "save time" simply through faster physical movement. Faster physical movement without thought, without reduced steps, without analyzing your operational process will ultimately lead to rushing. When we rush, we skip key steps, and we fail to complete appropriate size-ups. This dangerous combination leads to mistakes which ultimately make us less than or completely ineffective, which increases the likelihood of injury or death for both ourselves and the public. At the very least, mistakes ultimately cost time as we must now correct course.

BEST CHANCE FOR THE BEST POSSIBLE OUTCOME

If we truly buy in to the heading above as our mission and we embrace the data that is currently available to us, then there is simply no other acceptable approach than to measure all phases of our operations down to the individual second. If we can acknowledge that every second truly does matter in the outcome of the unprotected civilian (and for our own well-being), then we have no other option than to evaluate all factors that contribute to speed and efficiency in the emergent setting, including our fitness, equipment, apparatus design, training, and our attitudes.

When we embrace the impact of time on the fireground and we embrace a holistic approach of chasing every inefficient second in operations, then, and only then, can we begin to defend our actions as those that provide the public with the best chance for the best possible outcome.

THE SEARCH: A CHIEF'S PERSPECTIVE

SCOTT THOMPSON

As you read through this book, I ask that you stop for just a second and think about three things. First, I ask that you think about why you got into the fire service. Was it because others in your family were firefighters and becoming a firefighter is what you were expected to do? Were you bit by the bug at an early age, and never grew out of the dream. Was it because of what being a firefighter in the American Fire Service represents? Next think about the awesome responsibility that is bestowed upon firefighters. During your career you will have many opportunities to make a difference.

You will save lives either directly or indirectly. Finally think about the most important people in your life... your kids, spouse or significant, sibling, or parents. As you read this, think about what you want for them in terms of fire protection. Are you the firefighter you want showing up for your family? Based on your current level of influence in your organization, are you doing all that you can to build the fire department organization that you want responding to your loved one? If not, why not?

MISSION DRIFT

Over the decades, but specifically in the last 15 years, the American fire service has experienced mission drift. So-called progressive urban and suburban fire department organizations have initiated or tolerated a narrative that devalues the importance of a rapid search for fire victims. Starting with the removal of Tot Finders, the introduction of NFPA 1500 – Standard on Fire Department Occupational Safety, Health, and Wellness, and the latest search related acronym SLICE-RS (Size-up, Locate the fire, Isolate the flow path, Cool from a safe distance, Extinguish, Rescue, and Salvage), and the mindset that because fires are down - a commitment to search is not a high priority, are among the reasons for this shift in thinking. While I do not believe any of the above were intended to allow fire victims to die, the various interpretations and implementation of NFPA 1500, Two-in/Two-out, and SLIC-RS in fire departments across the Country has resulted in confusion and indecisiveness when it comes to prioritizing the search. What we have not heard much from over the same period of time are advocates or standards that focus on fire victim survivability and recovery.

SETTING THE TONE

As Chief of Department, I have a considerable amount of influence over the organizational culture, what the culture values and priorities. This is a responsibility that I take seriously, and that I manage daily. The chief, often through the operations chiefs, deputy chiefs, and shift commanders sets the tone when it comes to prioritizing people, in this case, firefighters and fire victims. Through formal directives and casual firehouse conversations, I am continually sending a message lending insight into my priorities and operational philosophy. Statements such as "fires are down, we are an EMS department that occasionally goes to fires, and it's not our emergency" send the message that being fire ready and having a culture that values fire victim survival and recovery, are not a priority.

We must never forget to consider all stakeholders when considering policy or guidelines on search. The safety of firefighters as well as fire victims must both be taken into serious consideration when establishing fire ground priorities. I have two sons that are firefighters in busy fire companies. As the chief of department, I am committed and driven to build a fire department response system that I would want my sons to work in, and that I want going for my five grandbabies. I often wonder what risk adverse fire chiefs are thinking about when they are ok leading an organization that does not meet the standard for what they expect for their family. How and why do so many fire service leaders casually write-off fire victims instead of digging in and building a search culture to assure no one gets missed? The formula to success is simple. Hire people with solid character. Provide quality daily training. Promote courageous, competent, tactical thinkers. Establish an aggressive response and deployment standard, and let firefighters be firefighters.

OPERATIONAL PHILOSOPHY

So just how can the fire chief influence search in their organization? The first thing that needs to happen is the leader of the organization needs to do some serious soul searching. The organization looks to leadership for direction. This direction comes in many forms. For the purpose of this topic, we will focus on the mission, operating guidelines, standard operating procedures, and operational philosophy that clearly outlines the commitment to risk acceptance and risk avoidance. The operational philosophy of any organization or firefighter within the organization, as defined for this subject as a theory or attitude held by a person

SECTION 1: MINDSET

or organization that acts as a guiding principle for action.

When we talk about searching a fire building, we have to reference philosophy. Why? Simply because there is not sufficient data to support the mindset that conducting a primary search is too dangerous. We are just now getting reliable data that highlights fire victim location and survivability in relation to time. The operational philosophy which often becomes policy, provides formal or informal insight on the priority of the search. The following are the things that decision makers need to consider when applying the mission and writing policy on guidelines that govern search. The failure of the "leadership team" to clarify the following places members in a position of higher risk and creates an environment of inconsistency and possible negligence.

SEARCH CONSIDERATIONS

COMMITTMENT

Training and the commitment to mental, physical, mechanical, and procedural preparedness is essential for search success and survivability. Leadership must support quality training versus quantity training for the sole purpose of checking off the monthly training to do list. Fire victims should never have to pay the price for lack of preparedness or a default pessimistic mindset towards fire victim survivability. Is search dangerous? Absolutely? Training and a commitment to preparedness with experienced supervision is the best thing we can do the make the search assignment less dangerous

PRIORITY

The priority of the search. Is the mission mindset to get the primary search done ASAP, or is establishing a water supply and knocking down the fire the first priority? We now have reliable data that gives us an idea of the amount of water needed for many single-family fires where fire victims are found, and the chances of fire victim survival in proximity to the fire room(s) and heat and smoke.

IDENTITY

Who will do the search? I believe everyone should be able to perform a rapid and effective search. However, I also believe in pre arrival assignments, and the separation of engine company work and truck company work. Rehearsal, in training or on actual events increases the chances of success.

ADVANTAGE

To give fire victims the best chance of survival and recovery, and to place firefighters in the best position for search success and survival, allow for "search specialist." Crews that have the primary responsibility for search can develop a search mindset. This allows them to be the most recently rehearsed at all things search. Having designated search crews gives firefighters a mental advantage. They can prepare for search success at the beginning of each tour, and they can go through a mental pre-arrival checklist while enroute. Firefighters are deprived of mental preparation when they have no idea of what they will be doing until they are staring at the problem. We can do better, not easier, but better.

> **RISK MITIGATION**
> Will the primary search be allowed to be performed without a handline? There is no question that searching with a charged hose line in hand slows the search and splits the attention and situational awareness of the search crew. There is also no question that searching independently of a line may have a negative impact on firefighter safety. I personally believe that with a quality training program and people assigned to do the search that I can justify the search crew working independently of the fire attack crew. I mandate that a charged line be in place and in the area of the search, but not with the search crew. This is a trust and training issue.

CLOSING THOUGHTS

Of course there are many more considerations from an operational perspective. The question is, as the Chief of Department, are we setting our people up for search success or failure? Just doing a search is not indicative of search success. Are unjustifiable fears and personal biases creating a culture that does not value an aggressive primary search? Should the unprotected fire victim suffer the consequences of an unrepaired system that places little value on fire victim survival and recovery? We, the chief, set the tone and give the red, yellow, or green light to many operational tasks and tactics directly or indirectly based on our past experiences and our personal stance on risk acceptance and risk avoidance. What message is your philosophy sending to your members? Do you support it?

SEARCH CULTURE
FRANK LEEB

Dedicated and well-trained firefighters save the lives of those trapped or overcome by smoke and fire every day. This is accomplished by quickly employing two of the best lifesaving tactics: a systematic search plan and rapid water on the fire. When staffing allows, the simultaneous execution of suppression and search and rescue operations while not waiting until water is on fire has saved countless lives over the past decades and is part of our search culture. This search culture ensures we fulfill our promise to those trapped- that we will come for you.

After all, saving lives is our primary mission.

REDUCED SURVIVAL TIME FOR VICTIMS

The modern fire environment has made search and rescue operations far more challenging. The main reason is the ubiquitous presence of synthetic contents, which are now found in all buildings. Synthetic materials have more stored energy than wood products of the past, resulting in much higher heat-release rates. This means synthetic materials discharge heat more rapidly than ordinary combustibles. Consider that a single upholstered chair made of synthetic materials has enough stored energy (2 megawatts) to cause a flashover in a room and that this chair will burn much more rapidly than wood or cotton fiber-constructed chairs due to its high heat release rate. To make matters worse, new building construction methods and materials—along with thermal pane windows—retain more heat and smoke within the fire area, quickly using the available oxygen and creating ventilation-limited fires.

When firefighters "open up" the building by forcing entry and ventilating, this ventilation-limited fire reacts to the newly available oxygen supply.

Likewise, openings made prior to our arrival can accelerate fire growth, smoke production, and spread. These openings, if not controlled, can cause rapid fire growth and extension, making conditions untenable.

It's crucial to recognize that fires today develop quickly, producing thick, black smoke and soot that act as additional fuel, accelerating the time to flashover. This significantly reduces the survival time of trapped victims and limits firefighters' opportunity to safely conduct a search without an operating hoseline. In such situations, it becomes the responsibility of all firefighters to exercise caution and control all ventilation tactics, including doors and windows within the occupancy. Remember, a single uncoordinated ventilation tactic on the fireground can have severe consequences. This understanding empowers us to act responsibly and effectively in such challenging situations.

SIZE UP

Members must begin planning their search during their size up of the building. By observing construction features (type of construction, window layout, access and egress points, whether the building is built on a grade, etc.), members may gain insight into the floor layout inside a residential building. This may help determine the number of apartments per floor and the floor plan of the apartment. Before reaching the fire floor, conducting a survey of the floor below can help determine the interior hallway layout, apartment numbering system, floor plan of the apartment and closest means of access to a fire apartment. This information should be communicated between the officers and the incident commander.

Wind direction and velocity are critical size-up variables, especially when windows are open or fail. Direct or gusting wind may suddenly increase fire conditions and fire growth within the structure. As part of their size-up, members of the outside team must report wind conditions and their effect on the fire and smoke conditions to their officer and the incident commander.

PRIMARY SEARCH

Although search tactics have evolved with changes in construction methods and fire dynamics, our overall strategy never changes. To conduct an immediate primary search, followed by a painstakingly complete secondary search. We cannot always determine survivable space simply by looking at the structure from the outside. Many factors, such as closed doors, create spaces where victims can be found and saved. We must search these spaces as soon as possible.

A primary search is an immediate search for life. This search is rapid but thorough and systematic. The primary search will be influenced by the fire, heat and smoke conditions in a room or area, and may occur prior to the application of water. It should be done quickly, with due regard for members' personal safety. If entry can be made and a search conducted without the protection of a hoseline, members may attempt to locate and contain the fire area, and then start the search for life from this point.

If the area is untenable, move in behind the engine company as it advances and search from behind the hoseline. The search for life in the areas adjacent to or above the fire area/floor should start immediately upon entering the area, followed by a search for fire extension. Any extension should be immediately reported to the incident commander.

SECTION 1: MINDSET

SECONDARY SEARCH

The secondary search is a thorough and painstakingly complete search for life of all areas that required a primary search. In addition, the secondary search must also include the entire outside perimeter of the building and all shafts, basements, cellars, elevators, roofs, etc. Its purpose is to ensure that no possible victims are overlooked. Time is not as important as accuracy. A secondary search must be completed before any extensive overhauling of the fire area is undertaken. Additionally, a secondary search must be performed by a different company or members who performed the primary search.

DOOR OPENINGS

The entrance door to the fire area is not just a point of access or egress. Equally important, it is a ventilation point that creates a flow path for fire, heat and smoke. This ventilation point allows both heat and smoke to flow from the fire area and also provides a fresh in-flow of air at the floor level from the exterior to the interior of the fire area. The contents of today's occupancies burn quickly and become ventilation limited. These fire conditions lack sufficient oxygen to further expand the fire. The inward flow of air from the door opening may be enough to cause a rapid expansion of fire conditions and limit the time for an effective search. It is critical to the safety of the occupants and the members conducting searches, that we control the entrance doors until a charged hoseline is in position to operate.

Members must determine whether the door will remain open or closed while the search is being conducted. The determining factor is whether or not the engine company has a charged hoseline and is ready to advance. If the engine has a charged hoseline, the door to the fire area can be left open. A member of the ladder company can chock open the door to facilitate the advance of the hoseline. If the engine does not have a charged hoseline at the door to the fire area, the door should be closed after the search team enters. Positioning a firefighter at this door pending the arrival of the hoseline ensures the door is maintained and effectively manages the flow path while limiting the oxygen into the area.

VICTIM REMOVAL

When victim removal occurs before fire suppression, consider the thermal and toxic exposure to the victim when choosing victim removal routes. In this case, consider an egress route that does not pass the fire compartment along the

path of travel. This route may differ from that taken by the firefighters who commenced the search. Alternate removal pathways may reduce victim exposure to lethal toxic gas and thermal burns during removal.

Several factors can impact and influence this decision. These include inaccessibility, obstructions, exterior conditions and access points, window size, occupant size, removal to ground ladder, fire escape or tower or aerial ladder. Additionally, the time to task completion must be considered. Positioning a ladder to the third floor for a rear removal may be an

unrealistic option due to the time to task completion.

When removing an occupant, consider the victim's breathing zone in addition to speed. Maintaining a victim's head at a lower elevation during the removal may result in a lower exposure to potential lethal concentrations of gases. Consider the difference in the smoke layer and heat and gas concentration from one to three feet. The one-foot level will typically have lower heat and less lethal gas concentration and may be below the smoke layer. Additionally, longer removal distances amplify the effects of elevation and speed.

CONCLUSION

Locating and removing victims from burning structures has long been part of the well-trained firefighters' search culture. While our goal is to stretch hoselines and apply water as quickly as possible, searching for trapped victims while the first hoseline is being stretched is an integral part of that culture. This is because seconds count when fire victims are trapped—they depend on us.

They depend on us being ready. They depend on us being trained. They depend on us coming for them.

SEARCH AND COMMON SENSE
BRYAN LLOYD

Photo by: Jillian Stewart - iShootFire

PRIMARY SEARCH:

If you read the mission statement of any fire department in the United States, there will be something in there about protecting or saving lives. One of the many reasons citizens look up to firefighters is because they believe that when they call us, we will show up and remove them from whatever precarious situation they are in…period. There are many ways we can accomplish this task, but on the fireground most citizens are found and removed from harm's way during the primary search. The primary search can be completed ahead of or in conjunction with fire suppression operations and is done in a rapid, systematic fashion by maintaining a balance of speed and efficiency. One of the things we practiced tirelessly during my time in the U.S. Army was land navigation. The rationale being that the first thing we must do is get where we were going, or all our training would be useless. The same logic applies to a primary search in that we must "get where we are going' as soon as possible by applying all the information we have gathered through dispatch, previous knowledge of the structure, bystander info (layout &victim number/location) and a solid search size up. The primary search is often being completed in high stress or zero visibility conditions, so the goal is getting a thorough search of all occupiable areas completed as quickly as possible with a focus on checking known victim locations. With the information available from UL and specifically the Firefighter Rescue Survey, we now have actionable data that gives us verifiable information on where citizens are most often located by search and rescue crews. We know that victims are most often found in bedrooms during firefighter search and rescue operations (45%) and other common locations include hallways (9%), bathrooms (6%), and family rooms (16%).

I have heard and been involved in many debates on the topic of search. We live in a time where information is more accessible than ever before and between fire publications, social media and the internet there is a lot of information out there

Photo by: Jillian Stewart - iShootFire

and some of it is conflicting. It boils down to this… search is an assignment, and it is incumbent on the crew performing the search to decide what strategy they will employ and what tactics they will utilize to achieve their strategic goals based off actionable data, training, experience, departmental policy and common sense.

*Note: Speaking of the information that is out there if you are willing to look, there are many great resources available, and I wanted to share a couple that have personally changed how I look at searching a structure. If you are looking for resources to assist you or your, department the Clackamas Fire District #1Rescue and Search document and the information/ training put out by Insight Training LLC on Tactical TIC use are both worth checking out.

PRIMARY SEARCH:

- All members on the search team should attempt to stay oriented inside the structure regardless of your role in the search

- If searching without a line ensure a charged hoseline is between fire and search crew, fire room is isolated or at a minimum search crew knows location of fire and feels that a search is viable based off real time information

- If search locates the fire while looking for victims attempt to isolate the fire room and relay location to fire attack

- Talking during a search = less is more. (Don't be that guy who wants to constantly talk on the radio or hold a running dialogue while you're searching. Attempt to keep conversation to what is pertinent. You are listening for victims calling out/making noise and focusing on your search)

- Once victims are found keep them as close to the ground as possible

- Prioritize areas with highest likelihood of finding victims

- Prior to entry, get low and scan room looking for victims and fire location / utilize TIC if applicable

- If using a TIC during search, it should not be used while moving through the structure. Perform your scan, identify where you are going, lower the TIC and move to that location. Use the TIC for an initial scan when entering the structure, to keep track of your crew and to identify doors/windows that can be utilized as secondary means of egress for victim removal or emergencies

- Ensure all members on the search team are briefed and understand the plan. This isn't a long process and can literally be done while moving to the structure

SECTION 1: MINDSET

ORIENTED SEARCH:

One member runs the team and will coordinate the search team actions while staying oriented to his/her location within the structure and continually maintaining crew accountability. As mentioned earlier, everyone should attempt to maintain their orientation within the structure as well. The team lead can/will perform the following:

> ### ORIENTED SEARCH:
>
> - Stay at the door/threshold of a room and act as a waypoint for the member/s performing the search
>
> - If team lead has a TIC, he/she can quickly scan the room and relay pertinent information to the searcher
>
> - The team lead doesn't have to sit still. He/she can sweep/search the immediate area
>
> - Pay attention to both positive and negative interior conditions
>
> - Maintain visual or voice contact

SPLIT SEARCH:

This is a type of oriented search that splits your crew allowing twice the amount of space to be searched while drastically reducing time in the structure and improving victim survivability chances. This works extremely well when searching two rooms that are close together. The crew will split into two teams with Team 1 taking room on left side and Team 2 taking room on right side. Each team completes a search of assigned area then continues moving to the next room/area. The teams can move independently of each other as long as they are in visual or voice contact

VES / VEIS / WINDOW INITIATED SEARCH:

Like the artist formerly known as Prince, this tactic has undergone several name changes. It was initially called Vent, Enter, Search then was later changed to Vent, Enter, Isolate, Search. I have also heard people refer to the same tactic as Window-Initiated Search. Are these all the same things?? There are an estimated 29,000 plus fire departments in the US and it's probably safe to say there are different viewpoints from differing departments on what constitutes what, but at the end of the day whatever verbiage your department uses, all members should be on the same page and understand the tactics that are being implemented when these terms are utilized on the fireground.

> ### VENT, ENTER, SEARCH:
>
> - Complete search size up and locate known or potential area/s with highest probability of victims
>
> - Take window/make access
>
> - Sweep area below window/sound floor
>
> - One firefighter enters room, and one stays outside or both firefighters enter room
>
> - If applicable utilize TIC to locate door to room and immediately identifiable victims
>
> - Move to the doorway of the room searching as you go, hook and look then come back in the room closing the door behind you
>
> - Finish searching room bringing victim/s back to window and remove from structure

WINDOW INITIATED SEARCH

This is exactly what it sounds like. After a search size up is completed and it has been determined a window access point will offer the highest chance of success, crews will enter the room and start their search. If fire conditions do not allow them to leave the room, they will employ the same tactics as on a VES/VEIS. If conditions allow, crews can leave the room and continue their search through the structure ensuring they are staying oriented to their entry point the same as they would if entering through the front door.

WIDE AREA SEARCH

Remove the residential search mindset when performing this tactic. This type of search can be extremely disorienting and is a tactic used on large structures/big box stores that involves utilizing some type of anchor point (usually outside the structure or IDLH environment) that crews can tie into and proceed to cover a wide area. TIC utilization will greatly enhance/improve a wide area search.

> **WIDE AREA SEARCH:**
>
> - Secure rope/tag line to stationary object outside the structure/IDLH environment and move inward
>
> - Stop and scan at intervals as you move throughout the structure focusing on identifying interior "landmarks" and staying oriented
>
> - Members can leave a person anchored at the rope and search out from the anchor while staying in visual/voice contact
>
> - Monitor air consumption
>
> - When exiting structure tie off rope where crews stopped searching and follow it back outside. This allows incoming search crews to quickly get back to the point where first crew left off

STAYING ORIENTED DURING SEARCH

I have heard many conversations / arguments on the "how" when it comes to searching a room. Instructors will say "This search is the most effective here or that search is the most effective there". The most effective search is the one you practice and feel confident performing in an IDLH environment when conditions are not favorable. Some firefighters are going to feel comfortable coming off the wall and relying on their experience and training to perform a quick, thorough search with limited fear of "getting lost" while others are going to be moving slower and focusing on staying oriented to each other and the wall. Anchoring on a wall can be done while performing a left or right-hand search. We teach placing a foot or feet against the wall, stretching out as far into the room as possible while sweeping the area. Come back to the wall and move down a few feet and repeat the process. This is not taught to be a slow, laborious process but rather we are moving with a purpose. Two searchers going in opposite directions can speed up a search but that is dependent on room size as it may be more expedient for one firefighter to search the room alone. While you are doing an "anchored" search and come across a victim or potential victim don't get hung up on the "I have to absolutely stay on this wall". Let common sense prevail and know that when you move off the wall to perform a rescue or investigate something you can move right back to the wall. This doesn't mean leaving the wall, crawling over a bed, under a desk, getting disoriented then trying to come back.

Remember... common sense.

THE WHY OF ENTERING WINDOWS

JUSTIN MCWILLIAMS

Photo by: Clackamas Fire

Why do I always consider entering windows for search? Because the outcome of our civilians' lives depends heavily on TIME and conditions. Windows can often be the solution to both. When we reduce our options, we reduce their chances.

Windows expand our opportunities to initiate our search from more floors, sides and areas, which increases probabilities.

I'm sitting in a house that has 3 exterior doors and over 20 windows. Two exterior doors open into non-isolatable livable areas and one door leads into an isolatable room by a pocket door, which is extremely hard to locate or close in low to zero visibility with gloves on.

There are no exterior doors that enter into the area of the bedrooms. There are no exterior doors on the charlie or the delta side of this house. There are 10 rooms without exterior doors, but 100% of the rooms have at least one window. There are a minimum of 3 windows on each side of the structure, with 15' maximum distance between windows for opportunities to initiate a search. The layout is easily recognized from the exterior, with the garage on the left and the bedrooms on the right. The front door divides the house almost equally in half.

As I look at the neighboring two story house, the exterior doors vs window ratio is the same, with no exterior doors on floor two. If we don't take

windows into consideration for entering for search or egress for civilians, we are missing opportunities to improve civilian survival outcomes.

UNDERSTAND AND EXECUTE OUR PRIORITIES

The search crew's priority is to locate, confine and communicate the fire's location. Fire engines are often the first apparatus on scene, locating the fire and initiating fire attack. This implies the search crew's first objective has been completed, the search can now focus on searching for our unprotected civilians.

When we arrive, size up the structure and activities on the fireground as we approach. If an Engine Co. is on scene, is Fire attack entering through the main entry? Fire attack takes the main pathways to the fire and are covering square footage as they make it to the seat of the fire. Fire attack and their hose can slow the traffic to targeted areas for our search, or may hold up traffic in the entryway, hallway, stairway, landings and thresholds. This hinders our ability to perform our objective in a timely manner and in turn will reduce the survival rate of our civilians. At times we are able to see that a traffic jam is likely ahead and we size up windows around the structure that will allow us to create a detour, which improves our effectiveness.

When fire is cutting off priority areas to search within a structure, spending time behind the line and waiting for passage will take up critical time. When we continually size up the situation, we can recognize the issue early and divert around the fire, often by using a window.

Stairwells may be burned out, leaving us to decide to either bring a ladder to the interior, which we will need to maneuver around walls and furniture and lay it over the burned-out stairs or to throw ladders to the windows and access the upper stories from the exterior. A benefit of having ladders on the exterior compared to throwing a ladder on the interior, is the conditions that we will be rescuing our unprotected civilians through and the complexity and time it would take to rescue a civilian down an interior ladder.

SCENARIO ONE

We arrive at a female screaming that her two-month-old baby is upstairs in her crib. We have a targeted area to search, due to a positive report of an unprotected civilian:

Are you prepared?
Did you bring a ladder?
If not, will you go back and get one?

We have a decision to make, enter by window or enter by door. What decision are you going to make? The decision is not always based on entry, once we locate the two-month-old little girl, we must get her outside and to EMS. When assigned search, our mission is to reduce civilian injury and death, which could be the difference of always being prepared to have a decision to make of entering a window vs a door.

SCENARIO TWO

Or we arrive at a structure fire that we decide to target the bedrooms down the hall or upstairs. We decided this partially due to locating 45% of our civilians in those rooms 24 hours of the day. Civilians on floor two may be notified of the fire late and their egress has been cut off. Smoke rises and traps civilians on floor two, they have probably been occupying the deadly elements longer than a civilian on floor one, if they have been unprotected in a non-isolated area. When we have the option to search by windows:

- We increase our opportunities to enter into the room directly.
- Have a protected egress to the exterior
- Can decrease our time to locate
- Often decreases our chances of being disoriented
- Can decrease delays of navigating the pathways to the targeted area.
- Can go around other interior crews & fire, that would slow our progress
- Allows us to have an efficient multi-prong search

ELIMINATE DUPLICATION OF EFFORT

We are more efficient and effective when we cover square footage only once. Size-up, who has entered and where they entered the structure. When fire attack has entered the front door, another search crew has entered a door, is entering the same point and covering some of the same area what is best for our civilians' outcome?

Photo by: Clackamas Fire

Our search should be organized in a way that will minimize redundancy. Multiple entry points is an efficient and effective way to ensure that we spread out and cover our own areas. Even on a one-story ranch, with bedrooms down the hall, if we start at the front door and two search their own rooms down a hallway, we will have to backtrack to get towards the main body of the house. If we enter a window for our targeted bedroom, we will be able to be more efficient with our coverage and work from one end of the home to the other, without doubling back.

CULTURE OF FEAR

Fear of entering windows is strong within textbooks, departments and firefighters that are not educated on the subject. Fear of death is often the reason. Next time you have a conversation with death being the subject of searching from windows, bring the following knowledge with you.

Death is not the only influence that creates fear, but injuries aren't tracked and documented to the same degree of detail. The term "ves" has been used since the late 1900s and has historically been lobbied as a "dangerous" way to search. Prior to the acronym, it was seen as an assignment of opportunity.

February 3, 1989, Brooklyn, NY. The last LODD that I know of, that a fireman died entering a window to perform a search. (Firefighter John P. Devaney)

February 5, 1945, in London. The only other documented LODD that I could locate, of a firefighter searching from a window. (Firefighter Frederick Davies)(Picture by Reginald Mills)

Since 2008 up until the writing of this article, there have been 14 firefighters that lost their lives while performing a primary search in a residential structure. Four of the 14 were in Single Family Dwellings.

THE WHY IS FOR THEM

Why do we always consider entering windows for search? For an improved outcome of civilian lives. Increasing the number of opportunities of entering to locate and exiting with an unprotected civilian, due to impediments of other firefighters, equipment, visibility, navigation and fire— each of these can be the deciding factor of a child living or dying.

PRIMARY SEARCH VS EMERGENCY RESCUE MODE

MO DAVIS

Photo by: Jillian Stewart - iShootFire

First and foremost, I want to be perfectly clear that I believe our number one mission in the fire service is ensuring life safety. Today's fire service must get away from the 'me first' mentality and get back to doing what it has always done best: protecting and saving the lives of those that call upon each one of us during their time of need.

The United States Fire Administration reports annual fire service trends. In 2021, the USFA found that fires have decreased by 5%; however, the fatality trend was trending upwards by as much as 8% annually. Today's fire service needs to find a way to correct this unfortunate trend. The fire service must change its mindset back to being a VERY aggressive interior profession. We must be prepared with training, equipment and most importantly, attitude. This mindset must resonate and be agreed upon by every member throughout the fire service: from the highest ranked brass to the rookie walking into the fire station on their very first day.

KNOWLEDGE AND SKILL

Firefighting is a critical and demanding profession that requires both knowledge and skill when dealing with multiple aspects of fire suppression. Two key strategies placed into operation on the fire-ground during structure fires are "offensive attack mode" and "emergency rescue mode". These strategies are specific in their approach and priorities; thus, this article will attempt to differentiate between the two operations.

The most hazardous duty of firefighters is to rescue occupants from burning buildings. This activity has always entailed calculated risks and placing the victim's safety above our own. This calculated risk does not mean taking unnecessary risk or being reckless. It stresses the need for being well trained and approaching the search and rescue systematically. Often, search crews will be searching in the fire room, ahead of the fire room, and/or above the fire room with or without a hoseline. As such, it is extremely important for the firefighter to be well trained, experienced, and competent in his/her skills.

TWO INITIAL SEARCHES

There are two types of initial searches: the 'primary' search and the 'emergency rescue mode'. It is important for each member to understand when the appropriate time to deploy either of these two methods.

With every structure fire the fire service makes, the fire company must respond with the mindset that there are victims trapped needing immediate rescue. This places one in an aggressive mindset where they are prepared, upon their arrival, to operate in whatever mode that is needed: offensive attack, primary search, or emergency rescue mode. Good fire companies have this game plan well before they arrive on scene of any incident. Great fire companies put the game plan into actual action.

In the fire service, 'offensive mode' refers to the operational strategy where firefighters actively enter the structure to perform extinguishment. In contrast, 'defensive mode' is where fire crews attempt to extinguish the fire from the exterior. The decision to make an "offensive attack" is based on many factors with one major factor being the possibility of trapped persons.

Key elements of 'offensive mode' strategy include:

> **Interior attack**: Firefighters enter the structure with the main purpose of locating the fire, searching for possible victims, and extinguishing the fire.
>
> **Search and Rescue**: searching for and rapidly removing possible occupants who may be trapped inside the structure.

It is important to discuss in detail primary search in the 'offensive mode'. The search and rescue assignment involves proactive and determined efforts to locate and remove possible victims who cannot exit themselves due to fire, smoke, collapse, or any other life-threatening circumstance.

Primary search during a structure fire is a critical aspect of firefighting which requires a swift, rapid, and systematic approach to searching the structure. Again, remember one may be required to search beyond the hose-line. The primary search is being conducted at the same time as the hose-line is being stretched into the structure, the ventilation assignment is being carried out, and fire extinguishment is starting to occur.

The primary search involves entering the structure with a sense of urgency while adhering to established practices and training. The crew assigned to the primary search is looking for possible victims, searching for the fire, and assessing the overall situation to identify potential hazards. Firefighters use their training and experience to navigate through smoke, fire, and debris filled environments. Communication is a key component to the search and crews should communicate their findings.

The primary search starts close to the fire and works its way out and away. If necessary, it then continues up to floors above the fire. This method is used to remove the most endangered first. In today's fire service, fire crews may use specialized equipment, such as thermal imaging cameras, to help enhance the crew's ability to work more rapidly.

IT'S ABOUT PRIORITIES

The 'emergency rescue mode' is a firefighting rescue effort that prioritizes lifesaving rescues over fire control and extinguishment. When deploying this mode of operation, firefighters enter the structure knowing that there is a high probability of victims. The priority is to conduct a rapid search and rescue operation in specific areas where victims may be found. Note, the main difference between aggressive primary search and 'emergency recuse mode' is information received and on scene observations. Emergency rescue mode is a highly effective way to save lives in specific fire scenarios.

'Emergency rescue mode' involves entering the structure with a sense of urgency while adhering to established training methods and standard operating procedures. The fire crew must assess the situation, identify potential hazards, gather intelligence, and enter the structure rapidly to make the rescue. 'Emergency rescue mode' is a critical

operation that emphasizes rapid entry, search, and exit.

The 'emergency rescue mode' refers to a state of heightened readiness and a specialized response during structure fires where the intelligence, information gathering, or occupancy type dictates the need for emergency rescues. The rescue mode encompasses a set of procedures, training, and equipment put into operation to address specific rescue needs efficiently and effectively.

THE DIFFERENCE

Primary search begins within the fire area, adjacent to the fire area, (if needed) the floor above the fire area and concludes with the remaining areas of the structure. The 'emergency rescue mode' begins where the highest probability that a victim may be found.

As stated, our primary mission in the fire service is to ensure life safety. The decision to conduct a primary search or go into 'emergency rescue mode' is not a standard or universal assignment. The decision is based on a thorough size-up of the incident. Fire departments far and wide use standard operating procedures to make informed decisions on what operation best suits the current situation. Specific intelligence and circumstance lead to the correct decision when it comes to searching fire structures.

RECORDING OUR RESCUES TO DRIVE OUR MISSION

JUSTIN LORENZEN

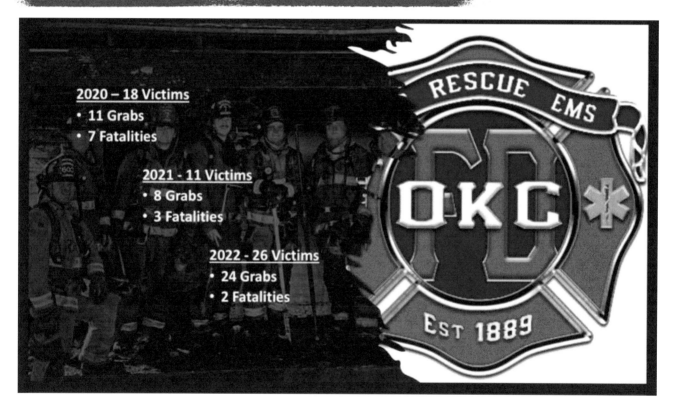

COUNTING OUR WINS

Traditionally, the fire service only captures civilian fire data on fatalities and injuries, but what about the ones we save, the ones for whom we make a difference? Why are we not sharing these outcomes? We weren't sharing them because we didn't have the information. This was a concerning issue for the Oklahoma City Fire Department. Year after year, we would hear of how many civilian fire fatalities our city endured, only to raise the question of "How many did we rescue?". Unfortunately, this question remained unanswered for years.

There was no doubt that OKCFD was making rescues, but as to how many, or any information on where, when, or how they were rescued, was not easily accessible, or even obtainable. This started the journey of finding a solution to capture this valuable information. This is where the Firefighter Rescue Survey (FFRS) comes in. The original proposal was to submit any rescue OKCFD performed to:

www.firefighterrescuesurvey.com.

Initially, this was not an option due to concerns of submitting incident information to a public online platform. The Firefighter Rescue Survey then became the mold to form the OKCFD Grabs internal survey, which ultimately led to OKCFD contributing to the national survey as well. More can be read on this process here, "The Oklahoma City Story: Tracking Grabs with Firefighter Rescue Survey", Fire Engineering.

DATA-DRIVEN CHANGES

The Oklahoma City Fire Department went live with their internal Grabs survey in 2020. This was a slow process and took some time to develop. Initially, when a rescue was performed, the crew was contacted directly, given the link to the OKCFD Grabs survey, and requested to submit a survey. Within a year the survey gained traction, and crews were submitting the survey without being contacted. Flyers were distributed to the stations that explained the vision of the survey and included a QR code that linked directly to the OKCFD Grabs survey.

During the time I have been with OKCFD, the only metrics that have been measured were our fatalities. Since OKCFD has started tracking our wins, changes in our operations have been made and we are getting to our victims faster. The first two years of rescue data comprised from the 43 recorded rescues has given us some valuable insights. From 2020 to 2023, the data show a significant increase in victim rescues to victim fatalities. This information, coupled with data driven changes in our search training and tactics shows a positive return on investment.

> In 2020, 54% of the total recorded rescues from OKCFD arrival on scene to arrival at victim were in 6 minutes or less, with a 67% survival percentage.
>
> In 2021, 62% of the total recorded rescues from arrival on scene to arrival at victim were in 6 minutes or less and had an 80% survival percentage.
>
> in 2022, 83% of rescues from arrival on scene to arrival at victim were in 6 min or less and had a survival percentage of 90%.

The data alone shows our firefighters are getting faster and more efficient at getting to our victims.

ANALYZE AND IMPLEMENT

This raises the question, "What are we doing to improve these times and outcomes?". By capturing and analyzing the data from our own rescues, this allows us to identify our strengths and weaknesses. We identify where we can improve and then develop a plan to accomplish it.

For example, OKCFD has adopted the "booster back-up" tactic, courtesy of Chief Curt Isaacson, from Escambia County Fire. Booster back-up is accomplished by the first arriving engine company bypassing a water supply and using their tank water. The second arriving engine company bypasses a water supply, attaches to the first engine, and gives them their tank water. This allows for fire attack to have 1000 - 1500 gallons of water to support the search while attacking the fire. The third arriving engine will then catch the hydrant and establish a permanent water supply for the remainder of the incident.

This allows OKCFD to put eight or more firefighters on scene to start search prior to, or simultaneous with fire attack in the first 6 minutes of fire department arrival on scene. Since

SECTION 1: MINDSET

implementing this, the data further reinforces that speed is our ally, and every second shaved could possibly be another life saved! In addition, the following contains other data points that stood out to us.

> 80% of all successful rescues (victims removed with a pulse or a pulse was regained prior to hospital arrival.) were made when search was initiated prior to or simultaneous to water on the fire.
>
> 86% of the total recorded rescues from arrival on scene until victim removed occurred in 8 minutes or less and had an 83% survival percentage.
>
> 66% of total rescues had reports of victims.
>
> 8% of total rescues reported that everyone was out. 67% of these reports were made by a resident of the fire building.

A LONG-TERM INVESTMENT

Like developing our internal survey, these operational changes didn't occur overnight. With multiple large recruit classes being hired during this time, along with a shift in culture within our training division, the needle began to move. Company evolutions were developed and drilled on, and having rescue data specific to our rescues contributed to the buy in. Recruits were being sent to the field with their probationary task books, along with new skills and tactics that some in the field had not yet been exposed to. Unintentionally, this intrigued incumbents in the field, and led to them seeking out the information on these new skills and tactics to be able to better evaluate the fire recruits. This was the beginning of OKCFD's search culture transformation. Skills and tactics backed by rescues made by our firefighters, along with the data from FFRS ultimately shifted our search culture.

ANYONE CAN DO IT

You don't have to be a large department that makes a lot of rescues to study and train on how/when/where/etc. victims are rescued. FFRS has data on 3000+ rescues that is 100% transparent and available to anyone at the click of a mouse. To date, OKCFD not only internally tracks our rescues, but submits them to Firefighter Rescue Survey as well. An interesting aspect of this was to see that OKCFD's data matched the FFRS data almost identically when compared in most categories. This further validated the information we were capturing and delivering to our members.

An internal survey may be the route your department wants to pursue, and that is fine. Start somewhere, so that foundation can be built, even if it's only one or two rescues a year.

The hurdle OKCFD initially experienced in submitting to FFRS was eventually overcame with the proper understanding of the information requested to submit a survey. The information submitted does not violate any HIPPA laws and all data is 100% transparent. FFRS also has a validation team that has the sole purpose of filtering through the data for any duplicates or inaccurate submissions. Each submission is carefully analyzed, and, if need be, the submitting department is contacted to verify the information.

As we push forward in our efforts and abilities to gather this data it is our responsibility to seize the opportunity to restructure and reform the antiquated ways of the past to better serve the citizens we have sworn to protect!

> *You can't go back and change the beginning, but you can start where you are and change the ending.*
> *-C.S. Lewis*

SECTION 2: PREPARATION

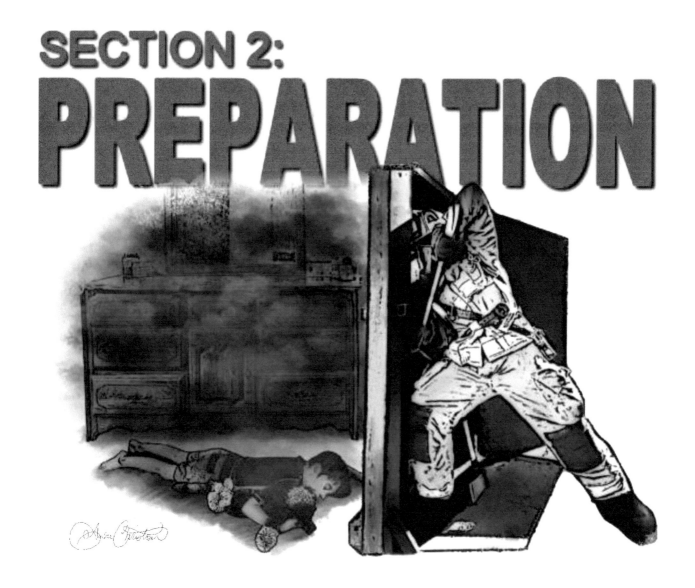

The articles in this section cover all the things you do to prepare. This is all about putting in the sets and the reps, it is about pushing back from the kitchen table, getting out of the recliner and getting out on the drill ground! It is about training! Some of it will be the mental aspect of how you prepare, and some of it will be the nuts and bolts of how to train. If it is about getting ready for the call, it will be in this section.

TRAINING FOR SEARCH
IAN BENNETT

When citizens pay their taxes or donate money to their local volunteer fire department, there is an unwritten contract between them and the firefighters who serve them. It states that if the citizen or their family is threatened by fire the fire department will enter the building, find them, and save them. To fulfill this contract there is no more important skill, besides putting out the fire, than the ability to conduct a primary search quickly and effectively.

Searching in zero visibility with an uncontrolled fire is not something that is easy or comes naturally. The ability to do it well usually improves as a firefighter goes to more fires but learning "on the job" alone is not acceptable. Failing to train for this critical task is a violation of the commitment between firefighters and the citizens who are counting on them. Thankfully most firefighters recognize this and are eager to learn and practice primary search skills.

OUR RESPONSIBILITY

In the real-world time, money, physical energy, and other resources are limited. This is true at every level from a senior tailboarder arranging a daily drill to chiefs designing department level training programs. As instructors it is our responsibility to ensure that these resources are not wasted on training that does not truly prepare our firefighters for the realities and challenges of primary search.

Practice for search can be broken down into a wide variety of individual skills starting from when the tones drop at the station all the way to when a victim has been successfully rescued and transported to the hospital.[1] Instructors can devote endless hours to such specifics as search types, location priorities, resource considerations, removal strategies, communication models, etc. etc. However, to effectively apply these techniques at a real incident firefighters need a solid foundation in two vital areas that are often missed, ignored, or done incorrectly in training. 1) Developing in firefighters the ability to rapidly read and understand the fire behavior inside buildings and how the actions we are taking will influence that behavior. 2) The ability, primarily by feel, to quickly and continuously gauge how far you

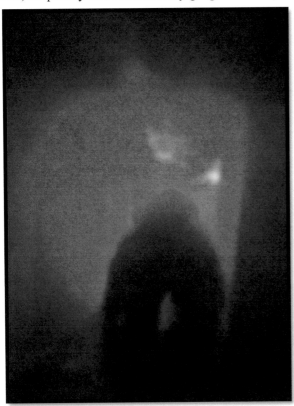

have gone, where you are and what is around you while under stress in low/zero visibility conditions.

These two foundational skills are critical for us to be able to make the more finite decisions of search specifics. If we don't understand where we can/cannot be based on the fire behavior and building layout, how can we accurately determine search priorities? If we can't count on crews to remain oriented in the building based on what they encounter, how can we accurately resource the search. Failure to build a solid foundation in these two vital areas will dramatically reduce the speed and effectiveness of any primary search even if the searcher is strong in the traditional skills/concepts of search.

Why the division into these two categories? It comes down to the challenges and risks of using live fire for training. First, NFPA 1403 places limitations on how live fire training can be done. This means certain important cues (the feel of a bed vs a couch vs a pile of clothes) and skills (dragging a live human) that need to be trained can't be done if there is live fire involved. Second, fire, smoke, and heat distract from the student's ability to focus on non-fire related search skills. This is fine if we are working with highly skilled students or doing a scrimmage style drill for unofficial or official testing purposes. However, it is very bad for cue recognition or physical skill development.

Learning to read the live fire environment is a skill that needs to be developed without the distraction of having to engage in the actual task of searching. Similarly, learning to gauge how far you have gone, and what is around you while in low/zero visibility, simply can't be drilled safely, effectively or in compliance with industry standards if live fire is involved[2]. Let's dive into each of these two foundational skills.

In the picture to the left the instructor, covered by a safety line, retreats after opening the door to the fire room. Students can observe the rapid change in conditions since they are not focused on the task of searching.

UNDERSTANDING AND PREDICTING FIRE BEHAVIOR

The ability to recognize what the fire is doing and predict what it is going to do next is vital to any search crew especially when searching without a hose line. Can I push to the next room? How long do I have to search before the fire makes it to where I am? Can I control the fire with a can? Should I take the person I have found back the way I came or stay

SECTION 2: PREPERATION

here? Without the ability to accurately understand and predict the fire behavior firefighters are only guessing. A strong foundation in reading conditions can and should be built during live fire training. Firefighters recognize this and regularly call for search training under live fire. The mistake that often happens is that when this training is developed the emphasis is placed on finding and removing dummies instead of reading fire conditions. Even instructors or students who attempt to focus on reading fire conditions will inevitably be distracted by the pressure to find and remove "victims" as quickly as possible. The search training then turns into crawling around in smoke trying to find rescue dummies that don't accurately represent what humans feel like in mostly empty rooms that in no way represent the reality of what firefighters will encounter in an actual building.

When live fire is involved, the focus should mostly be on creating a wide variety of fire, heat, and smoke conditions in different areas of the building and having students move through those conditions in different ways. The focus is not on traditional searching but on building familiarity.[3] The skill we are developing is the ability to understand quickly, on an intuitive level, what the conditions encountered in any given area mean to the search team. Visual, sound and heat cues are tied to conditions and actions developing the ability to "read" the environment. In the picture to the left the students are entering the building to conduct live fire training and can confidence. The focus is only on fire behavior and how the students can expect to influence it with a pump can. Smoke velocity, color and height are read at the door, students, guided by an instructor, take an intentional pause inside the door to feel heat, observe lift and roll-over, then the can is applied and the effect observed. A firefighter I know was the first person through the window on a ground-level VES where they were able to pull a victim out. The firefighter entered the window and encountered hoarder conditions with zero visibility and moderate heat. (The ability to evaluate the heat as moderate was heavily informed by his live fire training.) He attempted to isolate the room but instead of a door, found a stove. He realized that instead of a bedroom he was in the kitchen and moved to see if he could locate the fire. Looking out into the living room, the heat became much more intense, and a glow could be seen down low and

rollover coming across the ceiling towards him. He judged he would be able to knock back the fire with a pump can long enough to continue the search and did so successfully. The search was challenging in the hoarder conditions, and it was difficult to tell object from object. Conditions in the kitchen began to deteriorate again but a few moments later the Engine made entry through the front door and began knocking down the fire. A few seconds later the team found the victim and removed them. The firefighter stated that live fire training was key to his ability to read the heat and smoke conditions quickly and accurately, "it slowed things down."

REMAINING ORIENTED BY FEEL

The ability to quickly identify direction, distance of travel and the contents and layout of a room based on touch alone is not a skill that comes naturally especially when fully encapsulated in PPE and on air. Firefighters need to spend time moving around in search environments that accurately reflect the layout and material stuff that we may find in real buildings. They need a chance to feel different objects, flooring types, etc. with all parts of their bodies while wearing their PPE so that they develop a rapid recognition of what they are touching. They also need the chance to see how much space they can really cover in a realistic environment in various positions and search patterns. This builds competence and confidence in the ability to quickly and correctly identify where they are, what they are feeling and that they have covered all the space. This type of training also allows crews/departments to build solid plans and procedures to search common spaces because you can engage in high repetition low risk searching.

There are several common errors that happen when designing this type of training. The first is a failure to design search spaces that reflect typical room sizes and layouts. The second is not placing real objects in the spaces for firefighters to interact with. Durability is often chosen over realism. However, with a little time and creativity designing this type of search training is not very difficult. The prop shown on the previous page was built with OSB, 1x1 lumber, reclaimed doors and windows and donated materials and simulates a common house layout. The same can be done with acquired structures or even a room in a firehouse. Instructors can be used as "victims" so that students can learn what a real person feels like. This is important since people don't feel like most of the training dummies that we use.

For example, in 2022 a fire officer and his partner were searching a room above the fire where the family had stated a relative would be found. Conditions were zero visibility, and moderate hoarder conditions. The TIC was not giving a good picture, so everything was by feel. The crew went through the room multiple times without finding the victim. Only when the officer "kneeled on something soft but not soft like a pillow but with the density of a human." did they locate the victim. The officer stated he believed he had training scars from pulling out hundreds of training dummies made of fire hose or hard plastic and that this may have delayed his feeling the victim.

In order to perform search well there are many skills that must be practiced and honed. However, for firefighters to truly be effective these skills must be laid over a strong foundation, one built on a solid understanding of interior fire behavior and the ability to rapidly navigate spaces and identify objects by feel alone. Without this foundation the skills will crumble under the heat of a real fire.

[1] (Terry N. Wollert Phd, 2018)
[2] (NFPA, 2018)
[3] (Edited by Cannon-Bowers & Salas, 1998)

GROUND LADDERS FOR RESCUE

SHAWN DONOVAN

Photo by: Jillian Stewart – iShootFire

Let's be clear that using a ground ladder for the removal of an unconscious person from a building on fire is the second to last resort. Using rope would be the last resort, but still an option. Any other method you could think of would be better than using ground ladders; aerial, tower ladder, drag/carry out. Having a conscious patient does not change that fact. Sometimes, no matter how well you have prepared, the method of removal will be decided for you.

Advancing fire service ground ladder training starts with a serious look at our current training, which is generally inadequate to say the least.

CRAWL, WALK, RUN

Our hands on skills training usually progresses in the crawl, walk, run format with basic skills taught and then built upon, all towards an end goal. With nozzle work that end goal could be advancing while flowing a handline. With Ladder work it seems our end goal is "ladder is up". We are stuck at the walk stage. Two fire service legends Lombardo & Pressler teach the heart of truck work stems from two main goals: Search & Vent. We can come up with a few training goals based on that, such as ladder-based rescue/removal, vent a window, overhaul siding, window drop on an upper floor, removing unconscious people from an upper floor etc…

Dividing the walk phase of post fire academy ladder rescue training could be done into 8 or more separate skills that need to be performed together. We need to build our skills beyond what we are taught in drill school. Breaking down a ladder rescue into smaller components could involve the following areas that could be worked on individually:

1. Selection
2. Throw
3. Placement
4. Climb
5. Clear
6. Transition
7. Move/Lift Patient
8. Lower

Photo by: Csmeadphotography

Doing a simple internet search, you'll find areas one through three are beat to death, so I'll be brief. When in doubt go long. Yes, 75.5 degrees is the preferred climbing angle per the manufacturer. This is the angle that best balances climbing angle and strength. These ladders are rated at 750lbs with 4:1 safety rating (3k lbs.), they are obscenely strong. Keep in mind that ladders are tested fully extended with only about two feet of overlap, the longer the overlap the stronger it is. If your ladder is too long put the tip on the sill for support. Yes, tip under the sill is the darling of the fire service but if the angle of the ladder is too low it will demand a lot more from the butting of the ladder. Resting the tip on the sill at low angles sends the weight of the tip into the sill instead of down the wall, which would push the butt out.

CHANGE THE GOAL

We need to change the focus of our ladder throws. The GOAL of the ladder throw matters when placing it. Currently the focus is "Paint the building" in aluminum which I find to be an awful waste of manpower. Despite the excessive focus on creating a stable ladder for climbing the truth is these ladders are thrown with zero intention of ever being used, and it shows. Some have decided that they will focus on all manner of chock block, Halligan, webbing and rake to secure the base of the ladder. They focus on that instead of getting reps on the ladder to find how the ladder behaves when not solidly footed or to gain confidence with the ladder's ability to support them at less than "perfect" angles. Fire service ladders flex and we can use that to streamline our operations. Members and civilians who have decided it is safer to hang out a window than deal with what's going on inside a fire building have already cleared a big hurdle on the path to jumping, pick your priorities.

GET UP AND DO THINGS

Really being proficient while working from a ladder means it has to be practiced as much as throwing the ladder itself. The level of confidence required to work from a ladder without worrying about the ladder itself cannot be gained anywhere else but through experience. Get up there and do things from the ladder; use saws, break things, advance lines. Work through the rungs, off the side of the ladder and of course moving things over the tip. At the very least climb the ladder after you throw it.

Yes, working from the ladder is important but transitioning from the ladder to something else is an aspect that gets overlooked. Just like any other skill, it needs to be practiced before its needed. We practice that thoroughly in our academy, so it is generally not an issue later in the career. But every skill is perishable. Transitions to consider are through a window, onto a pitched roof, flat roof, porches etc.

THE TOUGHEST ASPECT

Getting away from the ladder itself for a bit, hands down, moving the patient is the one aspect of this whole operation that will humble a firefighter really quick. "We don't know what we don't know" could be the battle cry of moving an unconscious victim through a window, while you are on air in full gear. I will not pretend I have the answers for this. Guidelines would be to use practiced teamwork, use any mechanical advantage and a high point if possible, and use MMA style joint locks instead of trying to grab and move someone with your hands alone. The small muscles in your hands and forearms will tire quickly rendering them functionally useless.

Of all the considerations, moving the patient is the one skill that you cannot fake. Poor ladder throwing and low ladder use skill can be faked to a certain point. Members can be slow, reposition climb slowly and it will all look like progress is being made. But there is no faking the patient handling. It works or it does not.

One of the simplest skills that can be worked on is throwing a ladder with as few members as possible. Know your throw. Know how much ladder you can effectively handle, then make sure to practice in different scenarios such as hilly areas, around obstructions, over fences, down alleys, with obstructions at the foot of the building etc. When it comes time for the real thing, the more variety you've seen in training the better chance of successful outcome.

Photo by: Jillian Stewart – iShootFire

Do not work above where the top of the ladder is supported. Not having someone footing the ladder gets blamed so many times by the low hanging fruit committee. The real issue is usually the member standing on the ladder above the roofline. Having weight above the line of support turns the ladder into a class 1 lever instead of a bridge. Basic physics.

NEVER A TWO-WAY STREET

No ladder is a two-way street. If you see someone in a window, remember it's your duty to search that space. If you throw one ladder and the civilians get on it, your search will be delayed until they get down. One option is to try to keep them off your ladder until you get inside.

You could also throw another ladder right next to it. Having the second ladder with a second member right next to you will make a huge difference when trying to get someone over the sill onto the ladder. Sliding them down the ladder will be no prize either, having the extra hands will be needed. This needs to be practiced. A second ladder can also be used to get around a civilian coming down on their own.

Emergency removal using the ladder as a cantilever is also a viable tactic. If the patient is in the basement or first floor it's a matter of getting the ladder in, hastily securing them to it and sliding it back out on the rails.

Some other things to consider are to take out the handrails, if possible, on porches. Having people negotiate the handrail and a ladder can be challenging, if the balusters or the whole rail can be removed it'll make the whole operation smother. Then they do not have to climb over the handrail.

Some members will never get comfortable with "less than perfect" throws. This will hold them back. After you get comfortable with the flex of the ladder you won't be as concerned with the chock block for leveling or tying the butt off to something. Another option is to "corner" the ladder. If you can get the tip of the ladder just enough inside the window frame to not come out when flexing under load it will keep the ladder from sliding laterally. Practice these things. Practice with a belay if needed. Getting comfortable with the flex and settling of the ladder is a must if any work is going to get done from it.

DOWN IN THE WEEDS

We practice getting into every little corner of possibilities in so many aspects of the fire service, down in the weeds if you will. But not laddering. We never push the limits of what is possible or tolerable with our laddering and ladder usage. It needs to change. Any serious discussion about ladder rescues needs to start with ladder usage. Yes, moving patients is a major component of ladder rescues but it is going to be a tough sell to get someone on the ladder when the rescuer is scared too. To move ladder usage forward, beyond basic fire academy proficiency, we must get to a professional skill level an advanced level.

There is nothing wrong with basic fire academy level skills, but that should be looked at as a start not an end point. After the basics we move onto less than perfect throws, working from ladders, then onto working from less than ideally thrown ladders. We have to build confidence in the ladder so we can streamline our operations. We need to get to the point where being on the ladder is the easiest part of whatever we are trying to accomplish from it. Bottom line: Don't just throw ladders in training, climb them as well.

TRAINING AGAINST THE CLOCK
BASSEL IBRAHIM

Photo by: Csmeadphotography

THE CLOCK MATTERS

Our ability to respond swiftly and effectively to emergencies can mean the difference between life and death. Regardless of the type of calls we are responding to, Fires, EMS, Dive, or Hazmat… our biggest enemy is Time, and smart speed is the solution. In my humble opinion, there are 4 critical factors that are necessary to succeed in any portion of our job.

- Teamwork
- Communication
- Leadership
- Preparedness

To hone all four of these factors, training against the clock holds significant importance. In this short article we will delve into the preparedness component of these 4 factors of success, specifically touching on the importance of timed based training, exploring how it enhances decision-making, and ultimately saves lives.

Training against the clock provides firefighters with invaluable opportunities to simulate real-world scenarios in controlled environments. Real life emergencies escalate rapidly, leaving little time for hesitation or error. By setting time constraints during training exercises, we are forced to make split-second decisions, prioritize tasks, and execute them with precision. This simulation of time pressure helps firefighters acclimate to the stressors they will inevitably encounter during actual emergencies. It instills a sense of urgency and sharpens our reflexes, enabling us to respond effectively when every second counts.

THE ROLODEX

According to the Firefighter Rescue Survey data, the survival rate of victims drops roughly 14% every 2 minutes prior to us making contact. As far as the clock goes, we are already behind the eight ball. The fire scene is very dynamic, and some of the obstacles will be out of our control. To provide the highest chance of survival to our victims, we should maximize training on the controllables:

Photo by: Camera Worthy Photography

- Masking up
- Forcing doors
- Throwing Ladders
- Predetermined assignments
- etc...

The brain is a record of past experiences, and our reaction is the product of how we respond to said experiences. The more exposure we have, the faster our brain can pull the data and come up with a solution. So, the goal is to attempt to simulate as many "unknown" situations in training to create a rolodex in our brain.

We cannot predict the future, but we can attempt to create it, by introducing stressors like, time constraint in our training so that we are exposed to them prior to real-life emergencies. It is important that crews train on critical decision making under pressure as to not waste time on the fireground. I always say "imperfection, beats inaction". It is better to make a bad decision than not make a decision at all. At least we are making forward progress, if said decision does not work, we can reevaluate and redirect. Indecisiveness creates chaos and wastes time, putting a person in what I call "the Limbo State".

Repeated exposure to timed exercises improves firefighters' decision-making instincts, enabling them to remain calm and composed in the face of adversity. It allows them to avoid the Limbo State and remain effective and decisive. This mental resilience is invaluable during actual emergencies where clarity of thought and decisive action can save lives and mitigate hazards.

By repeatedly practicing these maneuvers against the clock, firefighters develop muscle memory and efficiency, enabling them to perform tasks seamlessly during high-stress situations. Repetition builds muscle memory, muscle memory builds instinct which in turns creates reflex, and ultimately saves time.

FOSTERING COHESION

Firefighting is inherently a team effort, requiring seamless coordination among crew members to achieve objectives efficiently. Training against the clock reinforces the importance of teamwork by emphasizing collective goals and synchronized actions. On my crew, each member has a specific "pre-assigned role" while also having

SECTION 2: PREPERATION

the ability to takeover/assist another member in their respective task if warranted. Each member must understand their role and perform it proficiently to ensure the success of the mission. By rotating tasks and positions during timed drills, we gain a comprehensive understanding of the entire process and develop versatile skill sets that will contribute to the overall team effectiveness.

In timed exercises, just like real life emergencies, every second matters, and any misstep or delay can impact the overall outcome. Thus, we learn to rely on each other's strengths, anticipate actions, and adjust strategies on the fly to overcome challenges swiftly. This collaborative approach fosters camaraderie and trust among crew members which enhances cohesion and morale. High morale increases productivity which in turn increases success. Training against the clock, fine tunes our ability to communicate effectively, delegate tasks, and coordinate our movements to maximize efficiency.

Effective decision-making is paramount in firefighting where split-second judgments can determine the outcome of an emergency. Training against the clock serves as a crucible for decision-making; it forces firefighters to assess situations rapidly, analyze risks, prioritize tasks, and formulate action plans under pressure. Furthermore, timed training helps optimize use of resources and leverage available equipment effectively.

EXPECTATIONS ARE KEY

When setting up training sessions, it is important to set expectations and clearly define the training type to your crews. Is it going to be a Skill based training session with some restrictions (i.e. "this is a Skill based training session, we will work on search techniques, you will perform a primary search of this structure while being blacked out simulating zero visibility, and you shall remain in a low stance while searching")? Or, is it going to be a Goal oriented training session without any restrictions, (i.e. "this is a Goal oriented training session, you will perform a primary search of this structure, and you may use any tools or tactics you choose") which in this specific scenario, crews may choose to use their flash lights, or Thermal Imaging Cameras. They may also choose not to search in a low stance if visibility permit.

Photo by: Camera Worthy Photography

CONCLUSION:

In conclusion, timed training plays a vital role in preparing firefighters for the challenges they face in the line of duty. By simulating real-world scenarios, fostering teamwork, enhancing decision-making skills, and ultimately saving lives, timed training equips firefighters with the tools and mindset necessary to excel in high-stress environments. We rely on our training, experience, and teamwork to protect communities and ensure public safety. Therefore, investing in comprehensive and rigorous timed training programs is essential to sustaining the effectiveness and resilience of our workforce.

THERMAL IMAGING TIPS FOR SUCCESS

ANDY J. STARNES

Photo by: Insight Fire Training

Firefighters face innumerable challenges in their daily service delivery. The challenges and dangers they face can be reduced and mitigated more efficiently using technological advancements through the implementation of tactical thermal imaging. As PPE has improved which allows for the better overall protection of a firefighter, it has also created a false sense of security where firefighters base their judgement of the overall thermal severity of the fire environment upon their senses. In other words, by the time a firefighter feels pain within the confines of high thermal protective performance gear, rapid fire development with thermal insult is imminent and an unprotected citizen is experiencing or has experienced thermal insult. The environment where firefighters work and execute highly skilled tasks is often one of zero visibility to very limited visibility. According to Firefighter Rescue Survey, 67% of victims were located in low to zero visibility. To assist firefighters with working more efficiently in these environments, we have developed the following tips for thermal imaging usage to ensure the successful implementation of these devices.

TIP #1:
CARRY THE TIC

Based on research that was conducted in collaboration with Firehouse magazine in 2020, over 50% of the fire service is failing to deploy the TIC until late in the incident on the fireground. The company officer is ultimately responsible for carrying a decision-making TIC to provide guidance and information to their crew's. A decision-making TIC's criteria can be found in the following QR Code. Please ensure that your organization is using an updated decision-making camera that meets or exceeds the criteria of NFPA 1801-The Standard on Thermal Imagers In The Fire Service. (whether it is certified or not). Antiquated/outdated devices are any device manufactured prior to 2013.

TICs can also vary in the following parameters: color/temperature correlation, field of view, refresh rate, resolution, temperature modes and thermal sensitivity. Therefore, to maximize clarity for the purposes of rescuing a victim and enhanced decision-making, firefighters should use a high-resolution decision-making TIC that meets or exceeds the 2021 requirements of *NFPA 1801: Standard for Thermal Imagers for the Fire Service*.

Situational awareness TICs aren't recommended for the purposes of tactical usage because of their lower resolution, slower refresh rate and smaller displays.

HOW ARE FIREFIGHTERS CARRYING THE TIC?

In carrying the TIC, many firefighters opt to use the standard retractable strap that is often purchased with the device by their department. Retractable cords are known for failure and can create entanglement hazards. The cords are designed to break at 60 lbs. tensile strength and are rated for 400 degrees Fahrenheit. They are stainless steel covered with a nylon coating. When these cords bounce due to the weight of the TIC, firefighters have inadvertently lodged their TIC in corners, on the edge of windowsills, and wrapped around the rungs of the ladder when making entry into a window. The carabineers that are often sold with these retractable cords are tension/spring loaded and will fail in the open position after repeated usage. This allows the TIC to detach itself from the firefighter during firefighting operations. Our recommendation that a firefighter use either a seat belt strap, leather strap, or some type of adjustable strap that goes over their shoulder after all PPE has been donned. The TIC will then slide up and down the strap on a large locking D-ring which allows them to scan and then place it back along their right or left side out of the way.

Photo by: Insight Fire Training

SECTION 2: PREPERATION

Photo by: Insight Fire Training

TIP #2:
BE PROPERLY TRAINED AND EDUCATED

In our experience, most of the fire service have had 2 hours or less of hands-on training regarding thermal imaging. 92% of firefighters surveyed revealed that it would be very helpful if TIC manufacturers would provide a training system in an electronic format. In addition, over 40% of firefighters have never heard of NFPA 1408- Standard for Training Fire Service Personnel in the Operation, Care, Use, and Maintenance of Thermal Imagers (2015). This standard outlines the proper procedures, guidelines, and recommendations for training on thermal imaging cameras. A word of caution to those who dismiss the NFPA standard. This standard uses the word "shall" quite often such as the following examples:

- 4.1.1 A TI training program shall be implemented.
- 4.2.1 The AHJ shall establish written policies for TI training that meet the requirements of this standard.
- 4.2.2 The policy shall address the training requirements for types of incidents where TIs may be used.
- 4.2.3 The training policy shall include an annual review of member competence in TI
- 4.2.4 TI training shall include practical evolutions, using TI.
- 4.2.6* The training program shall include both individual and crew training.
- 4.2.7 Members shall be provided with TI training and education before being permitted to operate TIs per AHJ.
- 4.2.7.1* Before new or unfamiliar TIs are placed into service, training and education relating to those imagers shall be provided for all affected members.

POWERING ON THE DEVICE

In addition, many firefighters who carried the TIC did not turn the device on until they were entering the structure. **A fire service TIC will require up to 30 seconds to be fully operational** upon pressing the power button. During its powering up process, the device is going through a process known as flat field correction. The screen may freeze every few seconds as it's completing its calibration process. Firefighters who fail to turn the device on until making entry often find out that the device isn't ready and a 30 second delay can feel like an eternity on the fire ground. This is also where many firefighters discover that their TIC battery is dead. Many departments either store their TIC directly on the motor cover housing on the fire apparatus or have a truck charger mounted in this area. This is problematic and can lead to battery failure. Lithium-ion batteries will fail to charge if kept in an environment over 112 degrees Fahrenheit. The company officer can place a spare battery in their turn-out gear pocket in case of battery failure.

NFPA 1408-and more are being quoted in litigation cases against firefighters.

Tom Zind summarizes it well in the following quote:

'Deploying the technology improperly, inaccurately interpreting the raw information it produces, and even failing to create an action plan that supports it can render results confusing at best or useless at worst.'

Photo by: Insight Fire Training

In summary, if a fire department has thermal imaging cameras in use; they **shall** provide proper training on these devices that meet or exceed NFPA 1408. And NFPA 1001-The Standard for Firefighter Professional Qualifications has been updated to require that all firefighters receive thermal imaging training. An industrial thermologist who works in the field must have a minimum of 32 hours of training on TICs to maintain a Level I Thermography Certification. However, firefighters often are placed in dangerous conditions without proper training, and they are expected to overcome this educational deficiency. This is gross negligence on the part of the fire department. Insight Training offers free resources to firefighters on our YouTube channel which you can find using the following QR code.

TIP #3:
LEARN TO SCAN PROPERLY

In previous years, it was commonly taught to use a 6-sided scan methodology with the scan process starting high. This starting point taught firefighters to look for imminent threat first. However, through our research and experience, we have documented that firefighters who start their scan high experience the following problems:

- An interior scan that starts high causes the device to immediately switch to low sensitivity. This occurs when the TIC detects a certain amount of heat and the details within the image diminish as it focuses on the superheated areas of the image. This can cause the firefighter to miss valuable information in the lower cooler areas where victims may be located along with the layout or floorplan of the structure.

SECTION 2: PREPERATION

- An exterior scan that starts high can cause the firefighter to focus in on heat in areas that are higher up (such as an attic space) without scanning below to see that the heat is emanating from a basement fire moving through the components of a ballon frame construction building.

The firefighter should always begin their assessment of any scene (exterior or interior) with their eyes and their training. If the firefighter picks up their TIC and begins viewing the environment first, they will find that the TIC greatly diminishes their field of view and could cause them to miss important information. As shown in the infographic, a firefighter's field of view is greatly reduced by the wearing of their facepiece. Once a thermal imaging camera is placed within their field of view, this could cause an optical loss of upwards of 140 degrees horizontally. The methodology that we recommend is a low to high method where the firefighter picks up their TIC in the gangster grip orientation, wipes

keep it in the cooler areas of the environment in the hope that it stays in high sensitivity for better image clarity.

Low scans require the firefighter to stay low which eliminates the issue of firefighters who stand up and attempt to use the TIC: A firefighter who stands up with a TIC will quickly discover that their facepiece and the lens of the TIC will fog up. This has been a contributing factor to numerous Maydays in 2023 where firefighters stated that the TIC "whited out." Newer thermal imaging detector technology doesn't white out due to heat. The screen will become white or washed out due to moisture build-up on the lens (the front of the camera has a germanium window that allows IR energy to enter). If this lens becomes occluded with moisture, smoke, debris, and particulate matter; the image will degrade or be non-discernable. A white screen can also be noted due to radio frequency interference, but newer models have an E.M.I. coating installed to prevent this from occurring. However, this has been seen in newer models that had a factory defect. Another cause of a white screen effect is due to the TIC switching from high to low sensitivity. A few of the current models produce a white screen briefly in between the change between high and low

Photo by: Insight Fire Training

the lens, and then begins their scan at the lowest area beginning opposite direction of the fire. We have listed several reasons for consideration of the low to high scan methodology below:

Low scans need to be conducted to look for life and layout first: By placing the TIC low, it will

sensitivity. If a firefighter is scanning too fast, they may experience this and falsely believe that the TIC "whited out."

Starting the scan low and looking away from the fire: By assessing the scene optically and determining the estimated location of the fire, a firefighter can begin their scan low starting away from the fire to look for cooler areas that may contain potential victims who have attempted to escape and now are near an egress point. Firefighter Rescue Survey shows that a certain percentage of fire victims are found near an entry/egress point.

Scans should be conducted slow: A firefighter that scans too quickly will find that the TIC may shutter or attempt to switch sensitivity modes/gains which can cause a delay or the image to freeze. We have documented hundreds of incidents where firefighters scanned too quickly and missed vital information such as the victim, the fire's location, and were unaware of changing fire conditions. If the TIC freezes while scanning, the firefighter should pause and let the device refresh and view what information it was attempting to process.

Sideways grip (gangster grip): Scans should capture the floor and ceiling in one view: By flipping the camera sideways, it increases the field of view dramatically. Carrollton Fire & Rescue Fire Department in Texas introduced this concept in 1999 by stating the following: "By turning the TIC sideways, this increases the overall field of view." This dramatically increases the field of view for the firefighter from as much as 9-15 degrees. A TIC with a 42X57 F.O.V. can now provide 57 degrees vertically by simply changing the orientation. This allows a firefighter to view the two most important areas during their scan simultaneously: the floor and the ceiling. Firefighters who use the standard grip in these situations often miss important details such as victim location, layout of the room and, in many cases, they miss the fire room due to scanning too fast. If the firefighter is looking up, the TIC's field of view is limited to the top area and often misses the area which is located 4ft and below. Therefore, they will often see the fire or signs of the fire in the form of convection currents, but they have missed the victim in many cases. When this occurs, many firefighters stay focused on the target (such as the heat source) and fail to scan the entire room such as the lower areas where victims may be located. We often assist fire departments in after-action reports involving TIC problems where we have learned firefighters have crawled over or past the victim. In other cases, due to improper scanning techniques, many firefighters have missed the fire room. My friend and mentor Battalion Chief Mike Mitchum summarizes this well in the following statement: 'The firefighter holding the TIC should take seconds to save minutes.' By scanning intentionally and not too quickly they can then direct their firefighters more effectively and efficiently. By slowing down their scan they can speed up their overall efforts. If a firefighter fails to stay low and scan low the TIC stays in Low Sensitivity due to the focus on the higher heat. This leads to a loss of detail and the potential for missing the victim is increased.

Once the scan has been completed, the crew leader shall communicate this information to their firefighter for task completion, and then put the camera down while moving. The crew leader should communicate the following information to the searching firefighters. T.E.D: This is the type of room (e.g. child's room with bunk beds), egress points (window and door locations), and direction of search. This can be communicated as follows: "A kid's room with a bunk bed to your 3 O'clock. There is a window to your 12 o'clock." At no point should a firefighter be moving (crawling or standing) while staring at the TIC display. By viewing the display while moving, they are increasing their risk of injury, increased loss of field of view (tunnel vision), and they are over-reliant on the device for orientation. If a TIC fails, a firefighter shall be aware of their surroundings so they can find a means of egress without the assistance of the TIC.

TIP #4:
PROPER INTERPRETATION OF THERMAL DATA

Firefighters who use TICs in these in a fire environment should be aware of how to properly interpret their specific brand of TIC. Each manufacturer is allowed to choose when their device shows colorization. This can be problematic and confusing as many firefighters are unaware of when this occurs and why. Firefighters need to learn their specific brand of TIC and when the device shows

SECTION 2: PREPERATION

colorization. Fire departments that have multiple brands within one fire department are also creating confusion for firefighters as they have to learn each model's various color/temperature correlations and other features. Firefighters also need to be aware that the absence of color isn't the absence of heat. Any white or moving gray/white areas are hot. These convection currents can be useful to identify the fire's location. The height of these currents can assist in determination of the overall severity. As this layer descends, the fire increases in intensity and develops into a full exhaust. Firefighters should be aware of the dangers of a full exhaust or a unidirectional flow path. Regarding temperature of these phenomenon, firefighters cannot substantiate the exact measurement as fire service TICs are qualitative in nature meaning they do not provide exact measurements. This brings us to our most important and misinterpreted area of fire service thermal imaging usage: **the spot temperature.** Many firefighters are still unaware that the spot temperature (or direct temperature measurement) number located in the lower right-hand corner of the display is not an accurate measurement nor is it a representation of the entire area. This is a small area of measurement that was calibrated in a temperature-controlled warehouse with a perfect black body temperature device. Many fire service TICs are off in measurement by as little as +/- 5 degrees Fahrenheit by as much as 41 degrees Fahrenheit directly from the manufacturer. Then when variables such as smoke, moisture, emissivity (reflective surfaces), and distance to the target all vary; the temperature measurement can be off by several hundred degrees Fahrenheit. For this reason and more, NFPA 1801-2021 standard removed the spot temperature from the start up application known as TI-Basic. It was also cited as a contributing factor in three line of duty deaths.

Firefighters need to be aware that these devices are

Photo by: Insight Fire Training

designed to detect heat and heat loss and only from surface temperatures. They do not see through objects, nor detect gases, or detect smoke temperatures accurately. For more information on this subject, please refer to our article: "The Danger of the Spot Temperature."

Important Note: Conditions affect image clarity and victim identification!

Firefighters have often been trained to locate a "white hot" victim by performing searches in artificial smoke while using heated manikins or live victims. However, the appearance of a victim through the lens of a thermal imaging camera will vary based on three conditions.

Background temperatures: If the overall background temperatures are greater than 200 degrees Fahrenheit, a victim who is still alive will present dark or gray depending upon their skin temperature and whether they are clothed or covered with debris (human skin temperature varies between 92.3-98.4 degrees Fahrenheit). It is important to note that animals that may be trapped in fire have higher core temperatures than humans so they may present differently as well.

Body temperatures: If the victim is alive and their skin is visible, they may present as dark or gray on the TIC display. If the victim is deceased and has been so for quite some time, they are now known as passive emitters. Therefore, a deceased victim may blend in with the background and may not be visible on the display of the TIC.

Type of Thermal Imaging Camera: If a firefighter is using an older low-resolution TIC or a situational awareness TIC with low resolution, their chances of locating a victim are diminished. Whereas a higher resolution decision making TIC may enhance their chances in locating the victim by providing important diagnostic information such as the layout of the room. When combined with the data from firefighter rescue survey, firefighters are equipped with a more effective search plan in terms of where to search. However, many thermal imaging cameras diminish image quality when in the presence of higher heat (or when the device switches to low sensitivity). Therefore, it is imperative that firefighters receive training on what these situations will look like before they are faced with the challenges of an actual victim rescue.

It's recommended that departments stage victims (rescue manikins) in three different temperature ranges to demonstrate how victims appear through the lens of the TIC. Departments that use heated rescue manikins will find that the manikin will show up white initially. A cooled manikin will show up dark initially and then gradually heat up and blend in with the background, as a human body will do in a fire environment. While there is no replacement for training with actual human victims for search and removal purposes, the limitations of using live victims in live fire training is currently prohibited.

In summary, this is a brief overview of Thermal Imaging Tips for Success. If you have questions or need other resources on this topic, please visit our website at:

www.insighttrainingllc.com

SEARCH: POSITIONS AND POISE
GARY LANE

Photo by: Camera Worthy Photography

WHERE WE STARTED

Over 25 years ago I entered the fire service and was given my first lesson on the topic of "search". I had no prior experience or preconceived thoughts on what that would entail. I did exactly as I was told, asked no questions, and felt like I was ready for our mission of "saving lives" after completing an hour or two of crawling on all fours, clinging tightly to either the wall or the guys' boot in front of me, all while swinging a pickhead axe around trying to feel something with it. And let me tell you- I was absolutely pumped about going to a fire and doing just that! It wasn't until several years had passed (I'm a slow learner) that I realized this textbook method was coming up woefully short. I must imagine that this experience, unfortunately, parallels many other firefighters around the country, past and present.

This above description isn't intended to place blame on any of those who oversaw our training at the time. Those guys were "the guys" for that department and had earned respect through hard work and experience. If anything, the fault was my own. The details of search and its refinement were most likely "lost in translation" at my completely inexperienced level of the time. Something that today, all these years later, I've tried my best to shorten the learning curve for anyone that trusts me to share with them lessons that I try to pass on.

Photo by: Camera Worthy Photography

One thing that I've tried to work on for years, are the various body positions we find ourselves in during movement through a house or structure when visibility is limited or non-existent. Standing, crouching, kneeling, lunging, crawling, and even a good old fashioned belly flop to "reverse snow angel". Anything goes in my opinion. I've walked and tripped over stuff more times than I'd like to admit. Crawled on all fours in a painfully slow manner. Taken a knee and scooted around lunge style. Poked with a tool and used my hands, stayed on a wall and come off the wall and been lost… the list goes on.

One thing that seems to have helped in minimizing problems, at least for me, is working through a combination of these movements, not just while in full bunker gear and SCBA at a fire or during training but incorporating similar movement patterns into my exercise routine. The higher my level of fitness, the more comfortable I feel while moving in awkward positions and searching. And since I'm not a big guy or winning any strongman contests, I need all the help I can get. The following are a few examples that have helped me, and hopefully could help a few others.

LUNGES

With weight or without, going from standing to kneeling is just about as basic as it gets. Don't take it for granted! It is easy to get off balance when we're tired, on uneven ground, or in poor visibility. Having the strength and mobility to do a basic lunge should be mandatory throughout our careers. SCBA adds weight and changes your balance. Do you wear them loose and sloppy or "high and tight"? Being in bunker gear adds an element of limited mobility due to how the gear fits. You may even want to consider this when sizing your next set of turnouts. I've seen both sagging pants crotch and floods cause movement issues.

BURPEES

The next movement that comes to mind is everyone's favorite: The Burpee. What a good time these bad boys can be during a workout. But again, the crossover physical body mechanics of going from standing, down to your belly, and back up again absolutely plays into a basic approach when initiating a search. Maybe a tweak you can add is going from a kneeling or lunge type position, and then dropping into the "burpee" from there. I'm sure you can visualize this in your head, or maybe you're already doing it. Kudos to anybody that's been trained on this. You're most likely ten times further along in your skillset than I was as a young fireman.

GET UPS

The last position I'll mention that can be practiced almost anywhere and anytime is something called the "Get Up" or "Turkish Get Up". Traditionally this is a move done with a kettlebell, but you can use any type of weight really (or no weight at all, as I have done many times) that works for you. I'll try and give a brief description here, but your best bet is to just pull up a couple videos on the internet, and then you should have a good idea of where I'm going with this. Basically with a weight pressed over your head, you begin to lunge to floor, post the opposite arm to stabilize in a pseudo tripod

SECTION 2: PREPERATION

position, the down knee side leg comes under and then out in front of you (several variations of this) and then your arm goes to elbow and then to flat as you lower down to laying on your back, all while trying to balance the weight or kettlebell pressed up above you. But you're not finished yet…. You still need to reverse everything in order and return all the way to standing once again. I've found versions of this to be a great way to get low to floor to look under a bed or smoke layer that doesn't result in my neck trying to bend at weird angles like when I've "belly flopped". Hope that makes sense, but I'm sure referencing some videos will clear it up if not.

STAY AWARE

Whatever you end up doing, or really the combination of things you end up doing while searching, I'd highly recommend trying to keep your head up and using your ears to listen as much as possible. This may require you to stop and pause for a moment, while trying to listen. Smoke is very good at muffling sound. Pausing for a second or two to listen has almost become a lost art with many in the fire service. Too much reliance on technology like thermal imagers and an overabundance of radio traffic has really created a "sensory overload" problem for us at times inside the fire building.

Whatever it takes, try to use all your senses (except for taste) and every so often take advantage of the pause, the start and stop rhythm of the fire scene, to get quiet and listen. Fire crackling, glass breaking, other firefighters around you, and the big one— the victims! A faint cough may be all you hear, or just some random rustling noise that is not coming from you. Something taught by many fire instructors around the country now commonly referred to as "Life-Fire-Layout" is a great start, with some of us adding "Call Out" as well. At any time during a fire, I may "call out" and pause for a second to listen. Something like "Fire department! Anybody in here?" And then listen for any sort of response. Give it a try next time you have an opportunity. It really should be a standard practice inside the fire building, especially if you're the first one through a door or into a room or making the next floor.

If smoke is lightest at the floor, that's where you'll want to get to when calling out, looking, and listening if possible. Hoarding conditions can change that. When encountering these types of conditions, it might be a better idea NOT intentionally burying yourself into several feet of contents, risking entanglement and the potential to have your facepiece pulled off by a random bicycle, clothes hanger or one of the four vacuum cleaners trying to grab you. These types of conditions will test your balance in a multitude of body positions as well. Slow down, communicate and be cautious of throwing things "out of your way" only to have them be in your way again a minute from now. I recently spent a solid 30 seconds getting my ass kicked by an empty suitcase at a fire with poor visibility and hoarding conditions. Thankfully, embarrassment was my only issue, and I quickly recovered.

Photo by: Camera worthy Photography

Another thing to think about ahead of time is what to do with the hand tool you're carrying. The most common tools that I use seem to be a halligan or 6-foot hook. Instead of sweeping the tool around a room like so many of us were originally taught, I prefer to leave it somewhere close by that I can still find it quickly, thus freeing up my hands to feel and move through bedding, clothes, linens, or other items encountered, in the hopes of not missing a victim due to lack of tactile sensation. A halligan or hook doesn't know the difference in what they make contact with like our hands do. If you choose to leave your tool someplace, try and have it off the ground, punched in drywall, hung on a door, in a corner, or some other obvious landmark that you can easily get back to and grab within a second or two if needed.

MAINTAINING COMPOSURE

Regardless of the fire's location, how much smoke is pumping out the eaves, or the presence of hoarding conditions, you must maintain your composure when searching. Because you will typically be operating without a hoseline in your hands, and many times be distant from the next closest firefighter, it is your responsibility to know your limits and pay attention to where you are, what you've searched, and where you're headed next. This is easier said than done sometimes. I have personally been turned around at numerous fires, and it was always when I stopped for a second, took a breath, got low and listened, that I found myself able to make the best decision on how to proceed. Sometimes this happened in a split second, and other times it took 10-20 seconds (or more!). Hands on training, repetitive basic drills, and conversations over coffee with the crew, can all play in to staying sharp between fires. You may even want to grab a beer with a trusted retiree off duty and ask them to share some of that senior man wisdom.

With so many resources available today to help educate our members and hone our firefighting skills, it's easier than ever to get started in the right direction with search techniques. Just remember though, with how fast fires can grow, spread and produce smoke, it's easier than ever to be moving too fast, become disoriented and end up needing someone to search for you! Practice and plan as much as possible ahead of time, and make sure to put in the physical work required to get the most out of your next search opportunity.

EFFECTIVE ARTIFICIAL STRESS

ROB RAMIREZ

As the modern fire service continues to evolve, today's firefighters are looking to adopt more effective instructional techniques. We long to leave behind the outdated and drawn-out PowerPoint sessions with their overflow of materials. This is the way the majority of topics are covered in traditional training programs. If we as a fire service training community start understanding the physiological benefits of properly introducing artificial stress during training, we could crack the training code. This would lead to successfully preparing firefighters of any tenure or experience level for actual stressful events. This change in our training methodology could ultimately improve their survivability, cognitive function, and physical performance under pressure.

HIGH CONSEQUENCES

The nature of firefighting requires that our members have the ability to function effectively under extreme forms of stress. Responding to emergencies, saving lives, and making split-second life-or-death decisions are all part of a firefighter's duties. When underprepared Firefighters are exposed to these extreme levels of stress it can significantly impact both their mental and physical performance. Most traditional training methods are often very linear. They focus solely on skill development and often fall short of preparing firefighters for the real psychological and

physiological demands of the job. In an effort to prepare for such harsh conditions, many of today's fire service instructors have unsuccessfully attempted to overcomplicate their training programs and ineffectively introduce the element of artificial stress during their classes.

THE HUMAN BODY

To properly inject timely artificial stress into firefighter training the instructors must first fully understand the strengths, weaknesses, and limitations of the human body. What we see, hear, and smell naturally drives our reactions and emotions during training and real-world incidents. This natural response to stress if left unchecked has the ability to overwhelm our senses and ultimately affect our fireground performance. Any form of prolonged stress can cause cognitive fatigue, reduced fine motor skills, and impaired coordination. This compromises a firefighter's ability to retain any new information or safely carry out any complex tasks. When the human body encounters stress, it activates the "fight, flight or freeze" response, releasing hormones like adrenaline and cortisol. These stress hormones prepare the body for rapid action: increasing our heart rate, diverting energy to essential muscles, and sharpening cognitive functions. In other words, we get really strong muscles in exchange for a really weak brain. Understanding how the body responds to stress can help you tailor your training programs to simulate these physiological responses.

THE BENEFITS OF PROPER ARTIFICIAL TRAINING STRESS FOR FIREFIGHTERS

- Enhanced decision-making
- Improved retention
- Better recall
- Increased mental resilience
- Improved stamina
- Increased physical endurance
- Enhanced focus
- Improved awareness
- Facilitates mastery
- Increased team competency
- Tighter team cohesion

MATCHING LEVELS

Introducing artificial stress in training is all about nature versus nurture. The form of artificial stress that you choose to use in your training must first match the skill level of your students and the training environment must be able to mimic real-world stressful conditions. This can be done through the use of noise pollution in the form of loud noises, timed tasks, realistic props, visual disturbances, or even virtual reality simulations. These types of training exercises require very experienced instructors and are not intended to be "a set it and

SECTION 2: PREPERATION

forget it" type of drill. When you use any form of artificial stressor to augment a drill it is imperative that every instructor participating in that drill is wearing the same level of PPE as the students and conducting ongoing assessments to monitor the student's stress levels and physiological responses during training, to safely adjust the stress variables as needed. The objective is to gradually accustom our firefighters to the physical and cognitive demands they will face in real-world emergencies, thereby reducing the impact and frequency of stress-

induced errors.

The impact of a training program must be continuously assessed through real-time metrics, participant feedback, and third-party audits. This ensures that the training is not just a singular event but a long-term investment in skill and knowledge acquisition. We understand that the most effective introduction of artificial stress in training is a gradual technique proven to help individuals cope and work through stress by systematically exposing them to it in a controlled environment. This process generally occurs in four phases:

PHASE 1: CONCEPTUALIZATION

Educating firefighters about stress and its effects on our ability to learn and perform. The instructors must define stress and our predictable human reactions to it. The first key for the student to manage stress during training will be in their ability to anticipate and identify it.

PHASE 2: SKILL ACQUISITION

Skills Acquisition and Rehearsals: When introducing the students to a new skill in training we must establish relevance and "marry" the new skill information with information that the adult student already knows to be true. As the student begins to connect the new skill with their old information, they start to form a new and more reliable understanding. Using this approach during early rehearsals will form their new baseline. Once the students begin to apply the new skills in training and struggle it is crucial that they are not left to fail to the point of panic, anxiety, or anger. This will completely inhibit the learning process and activate their fight-or-flight defense mechanism. If you notice that a student is starting to exhibit early signs of frustration it is important that you intervene and coach them toward a successful outcome. I like to tell my students during this phase, "Let's stop and take a step back to where life didn't suck for you". This allows them to take a step away from the chaos and start thinking about the new skill again since we understand that it's impossible for them to achieve success from a position of failure.

PHASE 3: CHUNKING

The method of 'chunking,' or the process of breaking down complex information into smaller, manageable units or 'chunks' has been popularized by many of today's most successful instructors. Chunking allows fire service instructors to introduce new complex skills incrementally. Firefighters can then master each chunk before moving on to the next, building both competence and confidence. Chunking also caters more to our typical adult learning habits by offering more flexibility. By breaking down complex tasks into smaller, manageable chunks, adult students find it easier to remember and internalize the information. This ability is crucial when our firefighters need to quickly recall life-saving procedures in a high-stress environment.

PHASE 4:
APPLICATION & REPS

What we do not fully understand, we do not fully possess. We cannot create effective fire ground behaviors by merely practicing or mimicking a skill. During this phase of Implementation, our students must first strip every 'chunk' down to its most basic form to truly internalize the action. As each moving part of the new skill is learned the instructors then increasingly add stressful variables into each scenario. The variables and speed of this process are one hundred percent student driven. It is important that each part of this phase is designed with the expectation of the students succeeding or learning. The most effective scenarios always provide an opportunity for success, but the possibility of failure.

METHODICAL PROCESS

The algorithm below is a basic example of a gradual and effective introduction to a new firefighting skill and the sequential incorporation of purposeful artificial stressors.

The instructor demonstrates the new skill, sets performance expectations, and provides clarity.

Begin skill and information "Chunking" repetitions.

Gradually progress to full PPE ensemble.

Create a metric expectation and begin to time the student.

Remove visibility.

Add physical exertion in the form of exercise before the drill.

Add passive smells prior to masking up (outdoor smudge pots, burn drums). Over the last four decades training research has proven that stimulating the sense of smell in training can support the improved memorization of information, stimulate information recall, and complement other human sensory channels used for learning.

Add sound pollution in the form of audible disturbances.

Finally, add OODA loop decision-making interrupters to encourage troubleshooting and contingency planning.

Repeat, Correct, Repeat

SECTION 2: PREPERATION

This methodical incorporation of artificial stress provides several physical and physiological benefits that significantly improve performance. As the human body and brain grow accustomed and resilient to these stressors it begins to expect and normalize those forms of arousal. This process creates both cognitive and physical improvements by enhancing memory retention, filtering out non-essential information, improving their decision-making, physical endurance, stamina, and overall mental resilience.

Ultimately your training goal should be to generate and maintain a more effective operational tempo while under stress. We cannot create effective modern fire ground behaviors while implementing a traditional linear training approach or by simply mimicking a skill. Having a deliberate training process creates an effective balance of mental and physical "progress capture", that is more readily

accessible under extreme stress. The arbitrary incorporation of artificial stress in training will only get you and your students in trouble. It is more important for us to know what is happening, where it's happening, and what's going to happen next, than to just robotically attempt unrealistic training skills that do not translate to real-world environments. Keep on training, commit to creating efficiency, and accept the fact that movement alone does not equate to competency. Movement without purpose is just movement and no form of practice has ever made perfect, practice only makes it permanent.

MASKING UP
MIKE BERNARD

Photo by: Camera Worthy Photography

1001, 1002, 1003, 1004…time ticks slowly as the Q screams and the wheels turn as we round the corner seeing a massive header in the sky, knowing, it is time to go to work. We have read the studies, we know the statistics, and we have prepared for this moment religiously. Air brake goes on, doors fly open, our bodies in motion and our eyes taking everything in. Adrenaline burns through our body with our heart racing as we try to take in the entire picture of what is in front of us. We are grabbing tools, ladders, hoses and everything else we need to implement our plan. We deploy to our respective sides of the building, specific jobs in mind and we go to work. 1040, 1041, 1042, 1043…time is still ticking, and we are fighting against a clock that will not stop until we make it. It is not our time to waste. We make calculated moves when fighting to get that time back. We gain access to the building while the clock keeps ticking… It's time to mask up and get inside. We don't know if anyone is inside or where they are, but we know we will search every space to give them a fighting chance. At this moment we could change the potential outcome of the victim with how fast we can get to them and to the seat of the fire. We put everything into motion with a great stretch, ladders went up flawlessly, and the door popped with a perfect force by the irons. We drop to a knee, prepared to make entry, everything contributed to this moment in time. We've masked up, pushing through heat and smoke, keeping this momentum going until we knock the fire and perform our search.

Sounds good on paper, doesn't it? Sounds like how we need to work on every fire. This needs to be how it looks when we close our eyes and dream of flawless execution. Or perhaps, did we get to the part about masking up and halt in our tracks because we weren't prepared for a skilled quick mask up? Seconds matter, our execution matters, and the small things we can do every day in training to be better prepared to win on the fire ground, 100% matter.

IT'S A LOT EASIER TO TRY IT THAN FIGHT IT

It is such a simple and small skill in this massive multifaceted barrage on the fire building, but it can add unnecessary time to get inside the building. There is a massive debate that circles around this skill and what the right or wrong way is. Are we masking up on the rig or are we masking up at the door? Let's be clear and get this out of the way right now, regardless of your feelings or which side you are on, can we all agree that we should possess the ability to operate our facepiece while we have our gloves on? It doesn't matter if we are a hard charger who lives and breathes masking up with gloves or if we are throwing our mask on while we are in route, we absolutely must be able to fix issues, take off/replace, and work all our straps and pieces of our mask and helmet with our gloves on. It's pointless to sit and argue about different ideals or who has a better way of doing it, instead, why not take the time to gain more depth in your skill set and work through masking up with gloves on? It's a win-win, if you don't use it as your preferred method at least you have put in the time to practice the skill and gain the ability to work through mask issues or put your mask on with your gloves on. If you end up practicing it and find out that it is your preferred method and put the time in to getting proficient at it, you are not only improving yourself, but you also can share it and practice it with others.

Photo by: Camera Worthy Photography

CRAWL - WALK - RUN

One of the biggest challenges we face when starting from square one is leaving our ego at the door and keeping an open mind. Skills that are new take time and require anyone and everyone to start at the beginning. There is no reason to eat the entire cake at once when you can take bites out of it over time until it is all gone. When we start exploring a skill such as masking up with gloves on, the best way to proceed, is one step at a time. No one is starting off their mask up journey with 10 second mask up times.

Instead of donning all your gear and running through reps that are fumbled and rough, just grab your mask, put it on, take it off, don't use gloves or anything else. Slowly become familiar with how far your straps pull, where the tabs are when it's on your face. Figure out the best way to slide it on without fighting it, where and which straps you can tighten

SECTION 2: PREPERATION

easily— finally creating a muscle memory to get it on effectively and efficiently every time. After you slowly improve and become comfortable with your mask, start slowly adding items like your helmet, your hood, your gloves, radio strap, coat, then finally your SCBA. Working slowly and methodically while adding one thing at a time will help you develop a system and process to do the skill the same every time you mask up. Creating this muscle memory is key to effective and successful mask ups at the door on each call.

STEP BY STEP

For those who are new to masking up with gloves on, one may wonder where to even begin? There are many ways to cut an onion and there may be some ways that are better than others. On the following page is a step-by step process that is a good starting point for new people to work with. Especially once you are comfortable with your equipment.

EQUIPMENT DIFFERENCES

One of the biggest differences that will alter how we mask up is our helmet chin strap length. A stock helmet strap that is not leather will usually be to short and nearly impossible to drop behind your head to don your mask during mask up. Having a leather chin strap that is longer will allow you to easily drop your helmet backwards to don your mask. Regardless of your chin strap, pushing it forward onto your arm and pushing your mask through the strap will allow you to keep your helmet off the ground while you mask up. This is a preferred method for people with short chin straps, but is not fool proof, work on this skill to develop your system to keep your times low without struggling to get your mask through the strap in front of you. Keeping our helmet on our person instead of it setting it on the ground especially in varying conditions keeps it with us and doesn't offer us the chance to lose it.

Find gloves, hoods, snap shackles, radio straps that all work for you and that you are comfortable with. Most of us are not blessed with the best and latest being issued to us at our departments. Make an investment in yourself and slowly add in personal equipment that works best for you and allows you to move quickly and effectively when doing mask up drills in training and in real life.

MASK UP HACKS

Now that you are starting to work past the crawl phase of standing and masking up in front of the rig, you can think about adding small tricks to improve your time.

A time saver to explore is taping your mask straps in tightened positions that not only allow you to don the mask but cut down on the straps you need to pull to finish tightening your mask. The only way to find out which straps you prefer to tape and tighten comes from putting in time on masking up. Find the straps that you can locate and pull 95% of the time versus having to search for and waste precious seconds on. The ones that you can't find or usually struggle with, you should tape, most people tape either two or three sides of the mask allowing for either both bottoms to be pulled or even just one.

Once you become more comfortable with your base skills, experiment with being able to throw your head net on without hands, letting it catch the back of your head and tightening it as it falls into place. Another experiment is to try and mask up without your regulator attached and see if that cuts down any time even with adding in the extra move of clipping in at the end of the movement. In some areas you will see over the hood mask ups as well as all pre tightened straps that allow the user to slide the mask on and keep moving without wasting time adjusting straps. You won't know if you don't try, and please don't knock anything you see until you try it or can offer an alternative way to do something constructively.

No matter the skill you are working on, having an unbiased party to watch you and critique you are paramount for your development. If we are training alone or on our own time, how do we accomplish this? Use your phone or get a video camera and record your mask ups for an uninterrupted and raw view of how you are doing. Sometimes there are small things you can see to improve on, or there may be things you would have never noticed without the video. Working with these tools and hacks at your fingertips, it is amazing how you can work on bettering your times and proficiency with masking up.

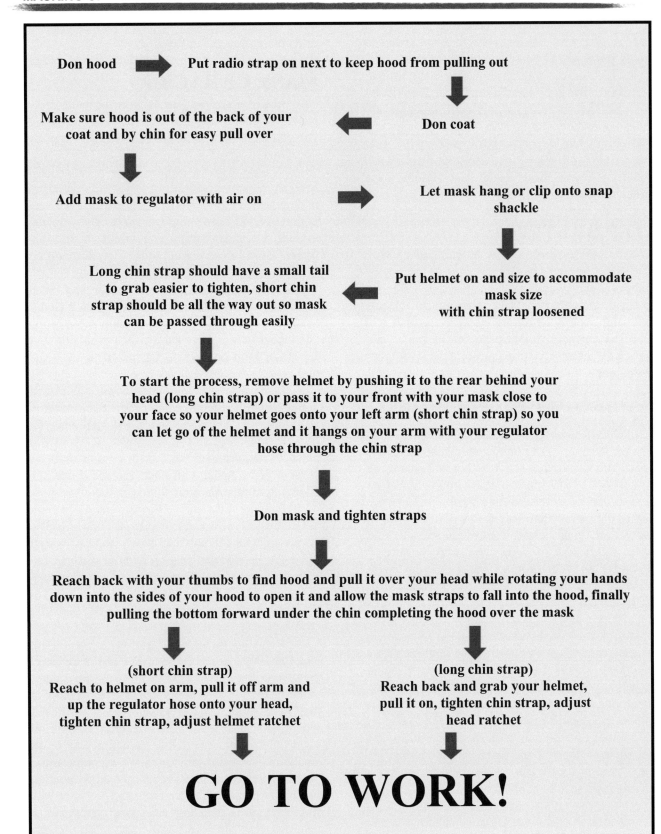

SECTION 2: PREPERATION

THE REAL WORLD

As we start to become comfortable with our progress with masking up, it's time we add in more real-life stressors to our process. Let's be honest, when we step off the rig on a real call it is drastically different from our calm environment at the firehouse in the apparatus bay. Get creative but realistic with what you are doing to challenge not only your mask up times but your stamina with muscle memory.

- Get out of the truck and throw all the ladders then mask up
- Get out of the engine and stretch to a door then mask up
- Force a door then mask up
- Do a 360 while masking up
- Run a mile then mask up
- Do burpees then mask up
- Turn the stereo up while radio traffic is blasting in the background

Photo by: Camera Worthy Photography

The list is endless and keeps going limited to only your own creativity. Pushing yourself to constant improvement, especially under demanding conditions will continue to sharpen your skills and challenge you in ways that you will see on the street.

BACK UP PLAN

Now we are really cooking along, we are masking up under 15 seconds, we have started with more complex drills incorporating other skills along with mask ups to make our skills sharp and practiced, but, what happens when our mask up does not go to plan? If we only ever plan for the perfect mask up every time we won't be prepared for when things go bad. When we look at masking up there are a variety of things that can go wrong and hinder our completion of the skill. Some things that we need to plan for are:

- Hood pulling out of coat
- Mask straps being fouled
- Helmet having an issue
- Regulator free flowing
- Seal not working

Our primary goal is to attempt to fix the issues with our gloves on, however, if that is not possible or it is taking to much time then we need to go to plan b. We start by removing one glove to fix the issue with one bare hand, if this doesn't work then we should remove our second glove and attack the issue with both hands. This can prove problematic if we drop our gloves or have our hands full already, but it is something we need to plan for and be prepared for when it happens.

Heat Straps are a great tool to add to our gloves that keep our gloves connected to our wrists if we must take them off. These wrist straps allow our gloves to hang without having to set them down and chance losing them in low visibility or environmental factors such as snow. If you don't have any special gear like Heat Straps, put them between your legs or under your arm while you fix yourself then get them back on and get back in the fight.

SUMMARY

We were able to stop the clock, we were able to get the fire out and complete a search, but how did we win? Did we fumble over ourselves and look like a soup sandwich at the front door? Or did we look like a well-oiled machine moving with purpose and executing our skills with precision? At the end of the day the responsibility of these skills and how often we practice them falls on no one else but ourselves. The standard we develop for ourselves today will drastically alter someone's life later in our career and it will be either a positive or negative outcome. Our data collection, scientific studies, equipment, and tactics are on the cutting edge and put us in a position to win every time those doors open for a call. We are given some of the best tools in the world, we have access to some of the best educational materials, and access to some of the best instructors in the world. None of it means anything unless we use it, it is on us to use it and make ourselves better because we must understand why we are here and why we must work harder every day to be at our peak potential. Masking up isn't the most glamours skill, it isn't the most in depth, or even the most complicated one, but it can mean the difference between life and death for a victim. Start with small obtainable goals and work at crushing them, eventually you can grow your goals and keep moving the bar for yourself higher and higher.

We don't fight alone, so don't forget about the people sitting next to you in the truck or engine. Don't be dismissive, cocky, or abrasive when others try this skill for the first time or if you get push back. We grow by working together and sharing the knowledge that we all possess to create the best outcome on the fire ground. We train together, we win together, and we need to be better for each other to leave this job better than when we found it. Remember why we are here, remember who we are here for, and remember why it's so important how we win…FOR THEM!

LITHIUM ION & SEARCH
PAUL SHOEMAKER

Structure fires are constantly evolving, and a new era of fires has emerged. The latest concern when searching for fires or victims inside structures is lithium battery fires. Understanding the potential fire behavior of lithium battery fires is crucial for firefighters to achieve the best outcomes. Firefighters need to comprehend why fires involving lithium batteries are becoming an increasing concern for rapid fire growth and firefighter safety.

OFF-GASSING

Recent studies have shown that lithium battery fires can exhibit severe fire behavior, such as flashovers, in a very short amount of time. One recent study demonstrated a scooter going into thermal runaway, and within a brief period, it started emitting gas at an aggressive rate. This process is referred to as off gassing. The off gassing can either initiate rapidly or be delayed, lasting for several hours. There is no concrete evidence to determine a consistent time frame for how long these devices will off gas before ignition. In serious cases with aggressive thermal runaway, the batteries can fill the entire structure with concentrated gas, leading to blackout conditions within one minute.

The interior structure may experience zero visibility. In this case the interior temperature can be the same as the ambient temperature. This gas that has just filled the compartment is not only

flammable but explosive in nature. When the batteries ignite, the temperature has been observed to increase from normal room temperature to over 1200 degrees within 2 seconds while all the gases in the room ignite simultaneously.

RAPIDLY EVOLVING CONDITIONS

The fire then exhibits signs of post-flashover, with a temperature decrease from 1200 degrees to around 475 degrees. This is where the fire transitions into complete flashover, causing pyrolysis of all materials in the room and causing a flashover, the temperatures climb back to 1200 degrees. This entire process, from 1200 degrees to 475 and back to 1200 degrees, takes approximately 30 seconds, leaving very little time to react and no consistent warning signs prior to ignition. This extreme temperature change exceeds the tolerance of firefighter gear, putting firefighters in a precarious position if they are not prepared for this rapid environmental shift.

When these gases fill the compartment, they produce multiple gases, with the most crucial being CO, CO2 and Hydrogen. This off-gassing depletes oxygen while simultaneously generating an explosive gas with the production of the hydrogen. The batteries create their ignition source once they reach their ignition temperature. Once ignition is achieved in the room or the area of concern, oxygen is introduced, allowing the lithium batteries to ignite themselves and all the gases in the structure. When the gases combine and reach the ignition source (the batteries), a chain reaction occurs, completing the fire triangle and leading to the instantaneous ignition of all the gases. This process results in a significant temperature increase and the explosion of all the gases, producing immense pressure. This pressure release could shatter windows or doors, breaking the glass and creating a pathway through the structure taking itself from an area of high pressure to areas of lower pressure.

SEARCH CONSIDERATIONS

With a basic understanding of the potential fire behavior associated with lithium batteries, it's essential to consider the scenario when searching within a structure. Whether you're first in on the ladder or engine assigned to the hose line for initial attack or assigned to primary search, the conditions could involve smoke (gas) obscuring visibility, accompanied by little to no heat. These conditions might create a false sense of security, leading to a tendency to venture deeper into the structure. Understanding the fundamental principles of fire behavior related to lithium battery fires is crucial. If you are conducting a search off the hose line, ahead of the hose line, or performing a VES, you might position yourself further within the structure without a readily accessible escape route. This could expose the firefighter to the explosion resulting from the ignition of gases, leading to disorientation and exposure to severe heat, potentially causing panic and further disorientation. Combining both issues greatly increases the likelihood of firefighters getting lost, injured, or overwhelmed within a minute or less under conditions of extreme fire and no exit route, compounded by disorientation.

Another concern with these types of fires is the thermal propagation throughout the battery pack. When these batteries enter thermal runaway, they eventually ignite, giving rise to the concerns previously mentioned. Once ignited, they can produce flaming projectiles, launching the cells throughout the structure, and in some reported cases, projecting them over 100 feet from the source.

Now, let's consider a scenario where you arrive after the lithium battery electronics have already ignited, and flames are emanating from the second-floor alpha-side window. Your assignment is to search the second floor. You enter through the front

door, ascend the stairs to the second floor, and commence your search. Now that we understand what these batteries may have been doing before our arrival, we must also consider the possibility of these projectiles landing in the polyurethane foam of a couch, behind a desk or in a blanket on the first floor. As firefighters continue their search, they may start feeling heat from behind them or on the first floor. Trying to retrace their steps out of the structure could expose them to heat and fire, blocking their path of egress through the front door and creating additional spot fires throughout the structure. It is important to incorporate the TIC throughout the structure looking for additional hot spots while searching through the structure. This can prevent the possibility of fire behind or below you, giving you a safer area to work in.

IN CONCLUSION

In conclusion, your approach, tactics, and, most importantly, your mindset must adapt to the rapid growth of lithium battery fires within structures. The standard search protocol can change significantly when dealing with lithium batteries, and careful considerations must be made as you navigate through the structure. You need to be vigilant for signs of additional fires in the area and constantly search for areas of refuge and alternative means of egress to ensure your safety amid these rapidly growing and potentially explosive fires.

THE PSYCHOLOGY AND PHYSIOLOGY OF SEARCH

RYAN SCELLICK

Photo by: Camera Worthy Photography

Primary search is one of the most physically and mentally taxing assignments on the fire ground. It is a true culmination of intelligence and physical strength, combined in one effort. The critical decisions involved in where and how to start the search, is just beginning of the cognitive load, involved in executing a highly skilled search. While most firefighters anecdotally understand the physical nature required to conduct a search, let alone a rescue, is significant; we rarely discuss how the physical load has significant impact to our mental capabilities. As the great Battalion Chief Rob Fisher once said; "truck operations are written in permanent marker, while engine operations are written in pencil." In other words, decisions made in waterless search can often be critical and unchangeable, because we don't have a hose-line.

GROSSMAN'S COLORS OF AWARENESS

Lt. Col. Grossman was a former Army Ranger, a paratrooper and West Point Psychology Professor. Col. Grossman is the creator of the expanded version of Lt. Col. Jeff Coopers "colors of awareness."[4] The work of these two men has been instrumental in understanding how the physiology of combat, affects an individual's ability to make critical and timely decisions. This body of work provides a critical link to the fire service, tying heart rate to cognitive ability. In this article we will break down the five colors and then discuss their critical application to the assignment of search.

- WHITE
- YELLOW
- RED
- BLACK
- (GRAY)

CONDITION WHITE

White represents a cognitive state of complete calm and free from outlying concerns. White is more psychological than physiological. Firefighters are rarely in the white mindset anywhere other than in the comfort of their own home. Heart rates that correlate with white should be in the normal resting range.

CONDITION YELLOW

In condition yellow, we move into a more hypervigilant cognitive state. Do you ever wonder why you are exhausted after a shift where you slept all night? It is probably from being in prolonged state of yellow awareness otherwise known as hypervigilance. In condition yellow, our heart rate increases from baseline to around 115 beats per minute (bpm). This is the stage where we have optimal brain capacity and function. We are utilizing our entire brain and still have full function of our neo-cortex or "decision making" brain.

Condition yellow is optimal for incident commanders or division supervisors, who need a high level of hypervigilance paired with a high level of cognitive abilities. Individuals that can maintain condition yellow on critical incidents, are typically those going to be removed from physical work, pain, or fear of harm.

CONDITION RED

Generally, condition red is going to be the most common physiological range a working firefighter will be in, at an emergency incident. In condition red, your heart rate range will be between 115 and 145. Condition red is the optimal physiological state for the average task-oriented firefighter, allowing you to still utilize your neo-cortex for optimal decision-making. In this range your motor skills, visual reaction time and cognitive reactions are all at the highest level of performance. However, starting at a heart rate of 115 bpm, we begin to see fine motor skills begin to deteriorate.

CONDITION BLACK

Above most firefighter's condition red, is the start of condition black. Studies show, condition black invariably sets in above a heart rate of 175 bpm. Condition black is a dangerous place for a firefighter to be. This is most commonly referred to as our fight our flight reflex. Condition black arises after the Amygdala (located in the mid brain), identifies something that is coded in our brain as a threat to survival, or is something we are unfamiliar with, that gets miss-coded as the same threat to survival. In condition black we lose all function of our neo-cortex (our decision-making brain), and transition completely to our mid brain or "lizard brain." In condition black, cognitive decision making diminishes, we experience vasoconstriction and shunting of our blood supply, we lose peripheral vision and can suffer from auditory exclusion.

CONDITION GRAY

You may have noticed there wasn't previously a color assigned to the heart rate zone of 145-175. This is because this is actually a "gray area" undergoing further research. While it should be noted that none of these heart rate numbers are hardened numbers that exist equally in all humans, it is believed that all humans experience the four-color zones discussed above, with black following red. However, research has found that highly trained and practiced experts can operate in condition gray. Condition gray is a zone above red that offers maximum cerebral and physical performance with minimal skill degradation when experts are pushing the envelopes of stress and exertion.

VERY APPLICABLE

Colonel Grossman's colors research was directed towards combat operations, but over the last two decades has become a fixed standard in law enforcement. However, there are no psychological or physiological differences between a suspect trying to harm you or experiencing the fear of you or someone else being injured in a structure fire. It should also be understood that these heart rate guidelines are in reference to a hormonal induced stress response, not an exercise-initiated response. While driving your heart rate up in training, does better prepare you for the real thing, it is not the same experience. In the remainder of this article, we will discuss how to use this knowledge to improve your search operation, as well as training to be successful.

When conducting waterless search, firefighters are pushing the edges of speed, efficiency, and cognitive processing. A gold standard of tactics in close quarters combat, is that you never move faster than you can shoot. The same holds true for doing a search in a hot, smoky, and dynamic environment. You can't move faster than your brain is processing things like; what is the floor furnishing, did I transition to a kitchen? Did I miss that window I saw in my size up, or am I off track? What is my air, what is my team's air? Is there water on the fire, are conditions improving or worsening, and on and on. I like to think of my searches as having a transmission with different gears for different speeds. I often relate searches to high level jiujitsu practitioners; they know when to be explosive and when to slow down, breath and process the position. For me, searching in a room is fast and dynamic but every time I complete a room, it's time to slow down, initiate purposeful breathing, and go through my checklist, air, heat, visibility and team location.

BREATHING

As Rickson Gracie describes in his book "Breathe,"[5] there is nothing more important than understanding how your breath connects with your brain and your body. I would venture to say that no matter what skills or tools you bring to your search, if you can't manage your breathing, your results will be limited at best. Whether your un-controlled breathing is sucking down a bottle, or driving you into tunnel vision, it is critical to control it. One of the only ways you can bring your neo-cortex (thinking brain) back on-line, is through appropriate breathing techniques.[6]

Photo by: Camera Worthy Photography

There are several techniques that exist to manage breathing, some good and some not good at all. None are more tried and true or as well researched, as the "four-box-breathing method." The four-box breathing method is utilized by special forces soldiers, swat operators and high performers around the world. As you breathe, you visualize a four-sided box, starting with a four second inhale, then a four second hold, then a four second exhale, then another four second hold, then repeat. Once you have perfected the practice, you can lose the visualization and apply this to any stressful situation. This breathing method causes you to engage your para-sympathetic nervous system, slowing your respirations and heart rate. It should be noted that when you exhale longer than you inhale, you further engage your parasympathetic nervous system, and when you lengthen your inhale longer than your exhale, you further engage your sympathetic nervous system.

Box breathing can be an invaluable tool on any type of emergency incident. While its popularity has increased, its specific application is not always expanded on. This can leave firefighters attempting to do box breathing during high intensity actions like searching, which is nearly impossible. This comes back to having search gears. When you are working hard and focusing all your attention on task level search, you will not be capable of box breathing. However, firefighters should establish in training, pre-determined checkpoints as discussed above; when to slow down, focus on breathing and bringing their heart rate into the lower part of the red zone. It is also important to understand that when you are working hard, you will likely need to start with a one second per side box and work your way to two, then three, then four seconds. Rarely do I reach a true four-sided breathing box in a hostile environment while performing search.

TRAINING

Training has always and will always be the birthplace of future execution. Understanding the psychology and physiology of your stress response is critical to applying it during training. Knowing what you feel like, when your heart is pounding and your respirations are running away, is crucially important to understanding when you need to slow down before you push yourself into a diminished cognitive state. Pre-determined phases of your search should include evaluating your breathing, fire conditions, available air and engine company progress. This should be implemented into your search training and practiced like every other skill. The firefighter who can push themselves into the gray zone while maintaining high skill proficiency will be an incredible asset to the fireground and to the people they are searching for.

---

[4] Grossman, Dave. "On Combat" 2008, Warrior Science Publications, China.

[5] Gracie, Rickson. "Breath" 2021, HarperCollins Publishers, New York, NY.

[6] Van Der Kolk, Bessel. "The Body Keeps the Score" 2014, Penguin Books, New York, NY.

THE DIRTY SIDE OF THE PRIMARY SEARCH

JEFF ECKERT

Photo by: Jillian Stewart - iShootFire

We all know, (or should know) that when we do a search on a dwelling fire, that we search EVERY area we can. And if we cannot search an area due to fire load, then we extinguish the fire, then search that area. We all want to be the firefighter "Making the Grab." It is exciting, it is glorious, it is what we signed up for! Find the people and get them out. There are many classes you can take, techniques you can learn, and tools you can buy to help you make the grab. Some firefighters go their whole career, in busy spots, and still never get a grab. That is just how it goes. Right place, right time, right position on the apparatus. In my 18 years in the Philadelphia Fire Department, I only have one legit grab of a civilian in a dwelling fire. That's it! But I have searched a lot of houses, of all sizes, and I have found my share of fire death victims. That being said, let's talk about what a lot of departments do

NOT teach... When NOT to remove someone, you found, from the building. This is a topic that is rarely discussed.

WHY THIS TOPIC?

So where does this article come from? What made me want to write about this? One night, my company, here in Philly had a first in job. Heavy fire showing first floor. Being that we are a Squad Company, we have an extra guy. During fire suppression we can have a guy perform a search. On this job while extinguishing the fire, they found a victim, obvious DOA. Burned down to almost nothing, Anther company, with a few newer firefighters came in, and even though the first in crew, full of more experienced guys told them the victim was dead, they still proceeded to pull out what was little more than a skeleton! Why??? Tunnel vision? Sure, but what about the constant talk about how "We are not doctors, we never assume that anyone cannot survive!" Yes, that is true, but that goes for, "Do we search or not," when determining whether or not someone has a chance of survival. We always search no matter what.

Photo by: Jillian Stewart - IShootFire

But can we declare if someone is dead? YES!!! Would you perform CPR on someone with no head? So why would declaring someone who is burnt to a crisp any different? And why is this important to know? For one, you have hurt the fire marshal's investigation by moving the body. You have now removed a body from the dwelling, and the medics are now stuck with them, on a street full of people yelling at them to "Do something!!!" Also, the family, if on scene now must look at what is left of their loved one. These are things to think about if you come across an obvious DOA. We stopped having cadets sing "No Glow, No Flow," because of the repercussions of not flowing water when you get in trouble but can't see fire. Water makes everything better, right? But we beat into everyone's head to "Get them out no matter what!" We never address when we should not get them out. And years of doing this have made their training scars, such as the story above and the next one I will speak about.

ARE YOU SURE?

Last year, we were 2nd in on a job with a perimeter company. We pull up to heavy fire showing from the 2nd floor rear, heavy smoke throughout. A lady was out front, screaming that her brother was still in the back room. Again, being that we are a Squad Company, we were moved on the Box to SOC duties and replaced with another engine. My job now was to help perform a primary search. There was no ladder company on scene yet. I meet up with the 1st in engine at the top of the steps. They are a very fast and aggressive company, so they already had good water on the fire. I told the officer (LT Mike) that I was going to search in front of them. I hit a left-hand search, find and sweep a bed, move to the center of the room and feel a person. There was zero visibility, but I think I feel legs and knees, in what feels like a chair. I move my hands up the thighs to the torso and my mask comes face to face with a charred skull.

I immediately yell out to LT Mike, "Yo Lieu, I found him, he is dead."

Lt. Mike responds with "Are you sure?"

I confirmed, I was sure.

He asked me again to confirm before he went over the radio, to tell the chief.

SECTION 2: PREPERATION

I confirmed again. This time saying, "I am looking at a skeleton." That was good enough for him.

BUT… when he went on radio to give the report, he was asked, "Are you sure???"

Once the 1st in engine finished fogging out, (hydraulic ventilation) not only could everyone see the damage to the man, but the chair I thought he was in was a plastic tote, that he fell in and was then melted to.

This is where I see a problem. Especially if we are looking for multiple trapped people. What good would pulling this man, melted to a tote, head burnt off, out to the street for all to see? We are wasting time and resources that need to be searching for live victims. We need to train our firefighters to be aggressive, yes. But we also need to train them to be realists. Pulling out a body in a zero-visibility dwelling is one thing. Sometimes you just cannot see and do not know. But there are times when guys can see enough, but just don't know enough to say, "They are gone." And move on. It is extremely tough to do that, but it is part of the job. We must save who we can and know when not to move someone. This will not apply to all fires. Probably less than 10%, to be honest. But when these mistakes are made, I believe they are detrimental to the fireground, the victim's family, and the fire investigation. Plus, it just looks sloppy and unprofessional.

WE START AT THE BEGINNING

So where do we start? At the beginning, when we have a sponge ready to soak up everything you tell them. Same principle as removing the "No Glow, No Flow" from the cadence. Teach them that when you search, try to search smart. Keep your cool. Be a professional. I have seen people drag a dog out because they came across that first, with confirmed people trapped. Hey, I love dogs too, but people come first. That is that tunnel vision, I am going to get a grab, be a hero mindset. I know, because I have thought the same thing on jobs. We can learn from our mistakes. That is one of the good things about making them. We LEARN!!! But if we do not change how we operate after making the mistake, then we will never improve and most likely repeat the same mistakes over and over. So, the next time you train with your company, or teach search in a fire academy, try to speak about when not to pull someone out. One day you may have to make that decision. Hopefully not, but you will be prepared.

SECTION 3: DECISIONS

This section is all about the decisions you must make once you arrive on scene. It is about how to process information and apply all those things discussed in the previous sections. From Size-up to strategy selection… from resource management to assigning tactics and tasks— this section focuses on all things decision making.

KEEP SAYING YES TO V.E.S.
Dustin Martinez

Photo by: Camera Worthy Photography

Vent-Enter-Search?
Vent-Enter-Isolate-Search?
Window Initiated Search?
Window Targeted Search?
Just… Search

Until the separate debate of standard terminology across the American Fire Service gets handled (spoiler alert: no time soon), the reality is that what other agencies call it holds minor significance as long as your department has standardized meaning to this essential operation. Terminology aside, the act of entering a window, finding a victim, and removing them to the exterior would appear to be the exact tactic that the untrained civilian would describe in what they believe a fireman should do on the fireground. And yet, controversy still exists when describing what should be a cornerstone of any search operation.

UNDERSTANDING VARIABLES

When discussing VES and the crew afforded the privilege to perform it, three main variables are present on the fireground: the victim, the conditions, and the crew.

Photo by: Camera Worthy Photography

To attempt to understand the current and projected state of the victim on the interior while approaching from the exterior is the nuanced art of survivability profiling. When determining survivability, we must look at all factors to have a truly enhanced understanding of the probability of survivability for the *unprotected respiratory tract*. This profiling would include the duration of exposure, the elevation of the civilian within the structure, their physical conditioning, their pre-existing medical conditions, isolation before the fire, isolation after they encountered fire conditions, the total dosage of toxic gasses they received, location concerning exhaust or inlet…. If that seems like a lot of information about the possible victim to process in a concise amount of time while attempting to decide the fate of the civilian you swore to protect, you would be correct. In fact, I would argue it is impossible to determine those things with any remote resemblance of accuracy. Now that you are spun up and agitated thinking I was going in a different direction, let me get to the critical point. It is about determining—

~~survivability~~
searchability

—when putting VES into motion.

SPEED IS OUR WEAPON

When determining conditions that would warrant the utilization of VES, the internal question the firefighter must ask is about speed. The measurable domain of time of opportunity to deploy the tactic decreases as the incident progresses. This shrinking target would include the ability of a search team to access the window and the interior hazard that is being stressed onto the civilian and their *unprotected respiratory tract*. With the advent of fire-based EMS, there has been a reliance and desire to apply the same protocol-based decision model to the American fire service. In an effort to embrace hope, change, and goodwill with the needle nerds (insert "EMS is most of what we do" response), I will gladly assist with creating this linear hierarchy of decision-making.

With your turnout gear, SCBA, and *protected respiratory tract, can you* enter that space in its current state and locate and remove a civilian? You can? FANTASTIC! Conditions allow you to search! Are you unable to enter the current intended target space to locate a possible victim? Then, make it searchable by modifying conditions (water) or find another place to enter. In the words of Chris Gilpin (and it pains me to give credit to such a hard-charging engine enthusiast), the nozzle team's job is to buy and hold searchable real estate. In the case that conditions do not allow you to enter, it is their time to create a buyer's market and encourage the engine to manipulate conditions to your advantage.

ONGOING ASSESSMENT

However, conditions can also change on the fireground depending on the time and the actions taken during your VES. The environment does not always deteriorate. When well-coordinated and decisive actions are taken, it may be BETTER than it was upon arrival. Although we are initially looking to occupy the first targeted area for access, we are constantly evaluating what areas past the first room of entry will allow for advancement past the entry point. If the intention is to occupy all searchable spaces, why would we give up operational momentum if conditions do not require us to do so? We also must remember that we may want to utilize an open ventilation point to our advantage after the fire has been knocked down to increase visibility while decreasing toxic gasses within the structure.

To the crew that has won the lottery and seizes the initiative to VES, this may be your finest moment. The task, location, and objective have aligned, and only you can intervene on behalf of the unprotected respiratory tract. And what a shame it would be if you were ill-prepared or ill-equipped to meet the moment. The embarrassment of arriving at the charlie side and seeing a window without the ability to attack it would pale in comparison to the fleeting hopes of the victim inside. For first-floor VES, always appreciate the importance of a shortie ladder. While the halligan or a police officer on all fours can serve the purpose, when meeting an elevated first-floor window (crawl space, slanted grade, mobile home aluminum ranch), a shortie ladder provides the easiest access and outside assistance with the victim unless a window to door conversion has been created.

If you have made it this far into the article and didn't anticipate a ladder for a window two stories or higher, take this advice to heart. Immediately stop reading, find your nearest community college, and enroll in their paramedic program. When throwing the ladder to win, the 2nd half of the operation means the victim will come down the same ladder. The ladder's angle should be set for victim egress, and high priority should be given to a second ladder to be placed alongside it to get multiple hands on a victim 13+ feet in the air.

UNDERSTAND TRAINING SCARS

In my initial training, I was instructed to always sweep and sound before making an entry. The intent of a quick z-sweep pattern is to make sure there is no victim immediately before entering and avoid a career filled with total remorse for pounding a tool into the unconscious person you are attempting to rescue. The sweep, however, is dependent on conditions and construction. Are you entering the first floor on a slab? If so, what are you sounding? Is this a training scar created by replicating a process regardless of the environment you encounter? Are you two stories above the fire, and your worry about fire in the basement requires you to sound the floor of division two? If you believe the floor you are going into needs sounding before entry, do so. If not, save time and get to work.

Photo by: Camera Worthy Photography

Photo by: Camera Worthy Photography

CONDITIONS DRIVE OUR ACTIONS

There cannot be a one-size-fits-all protocol for actions once entering the room. If conditions require you to close the door immediately, then close it. Priority should be given to extending past the room's door before isolating by looking down the hallway with your eyes and flashlight to see if any civilians tried to exit the room. If you find a victim in the hallway, bring them in and close the door. What if you locate a victim between the window and the door? If you can't quickly pick them up one with one arm and get them out the window, close the door. You would be doubling down on the known victim. What if you find a victim on the way to the door and one in the hallway.... Do you see how many variables there may be? If you still need the step-by-step protocol, please join your friend at Itawamba Community College's paramedic program.

Instead, there should be a reason behind your operational decisions. If you are attempting to create an area of safe refuge for extended victim removal, multiple victims, or conditions would rapidly deteriorate, close the door as quickly as possible. This will protect you from the ever-mysterious death phantom known as *the flow path* (UL, you owe me 20 bucks) and begin ventilating the room. You now have increased visibility to assist your search and provide a better environment for all respiratory tracts inside the room.

If water is on the fire, you may prefer to aid in ventilation and keep the door open. If conditions allow you to move past the initial entry point, move forward! You have now begun your next assignment, the ethical operation known as....search. Depending on your department, the incident commander may need to know that outside egress points can be established sooner if you find the victim two rooms over. Some safety chiefs wearing vests may raise concerns about secondary egress when you VES on elevated floors. Thanks to your ladder, you brought an aluminum exit with you, and the building gave you one with its staircase. Historically, we have never hesitated to send a search crew to the second floor through the interior stairs. But for some reason, the same crew climbing a ladder and dumping into a 14x14 bedroom now creates occupational chaos.

THE PENDULUM SWING

Fortunately, the American Fire Service's pendulum appears to have swung back in favor of the victim. It is odd even to think there was ever a commitment in the other direction, leaving the civilian to their demise and allowing cowardice to identify its pronouns as safe/ty. When actions on the fireground are initiated in the best interest of the fragile airway, the protected one will be safe and sound.

I want to be very clear about the intent of my contribution to this project. You must be proficient, ready, and reliable when the diminishing window of success appears on the fireground. VES is not new or groundbreaking. The tactic has been performed in front yards, from fire escapes, and on ladders. My beliefs are even repackaged forms of my mentor's teaching, and the information received. This article intends to display what I believe to be the most streamlined operation for entering, locating, and removing the civilians we swear to protect. The front door is always an option. It cannot be our only option and may not be our best option.

Across the country, brothers and sisters are performing VES, giving the chance of life an *actual* chance. It would discredit these same men and women, the American Fire Service, and the citizens we chose to protect if we were to believe that the aggressive and direct action of vent-enter-search is not a viable tactic. To quote Kevin Lewis and Keith Lawrence, "Say Yes to VES."

RULES OF ENGAGEMENT FOR SEARCH

JUSTIN MCWILLIAMS

Photo by: Clackamas Fire

When we are called to serve at a structure fire, we are to search all searchable space. When there are multiple companies or crews performing search, we coordinate our efforts to cover as much of the structure as possible in the most efficient and effective way. We will be educated and execute the "Rules of engagement for search".

RULE 1:
SIZE UP IS CONTINUOUS

When we are tapped out to a structure fire, we start our search size-up, what type of occupancy are we going to? Occupancy will typically assist us in forecasting the type of rooms, furniture and a rough layout. As we approach the structure, we size up the square footage, length and width, # of floors, possible fire location, smoke conditions, windows and doors to assist in the layout and understand our entrances and egress potential. This will be performed no matter if we are to enter a door or a window. As we move into the interior, we will continue to orient ourselves and size-up the interior. Fire location, extension, fire attacks progress, heat, visibility, confirming layout, types of rooms and searching all searchable space - no matter the way that we entered.

SEARCH SIZE-UP:

- Occupancy
- Era
- Square Footage/Number of Floors
- Layout
- Conditions
- Searchable Space
- Suppression & Ventilation Tactics
- Point of entry (window/door, floor, side)

RULE 2: LOCATE THE FIRE

The rules of engagement for search pertain to us no matter the floor, side, window or door that we enter. Operating inside a burning structure with an unknown fire location is one of the riskier times that we operate on the fireground. When a crew enters a structure, locating, confining and communicating the location of the fire is the priority. This will typically happen by either fire attack or the first search crew that enters the structure, if they are searching ahead of the line. This allows everyone to operate inside, with a "known" fire location. Search being light and fast allows them to locate the fire in less time and the ability to keep it smaller, by either knocking it down with the can or confining the fire to the smallest compartment possible. Communicating the location to fire attack, allows for an efficient stretch.

COMPANY SIZE CONSIDERATIONS WITH AN UNKNOWN FIRE LOCATION:

Typically, when a company comes off the rig with four and assigned search, the first two are the "inside team", with fire being their priority. This allows the "outside team" to target an area for people, enter a window or a door and perform a search.

Companies of two, that don't have the luxury of splitting two for the "inside team" and two for the "outside team", may still enter windows for search.

When fire attack is already on scene and located the fire, this allows the search assignment to immediately prioritize targeting an area to locate civilians. At times we arrive on a "dry apparatus" with no water, an unknown fire location and reports of a victim inside. We must make a decision to prioritize locating/confining/communicating the fire or prioritize the civilian that is being reported inside.

Photo by: Camera Worthy Photography

RULE 3: ISOLATE & VENT

We isolate civilians, ourselves and their belongings from the conditions. When we enter a room from the hallway or a window that is isolatable, we isolate when we have unwanted ventilation or introducing products of combustion, to protect unprotected civilians and delay extension and reduce their exposure. Isolating the room that the firefighters enter, is not locating or confining the fire but only closing off a small isolatable area within the structure. This rule of engagement assists in keeping fire as small as possible and with locating fire and confining it, allows us to search as much square footage as possible.

When we enter a window for search, we typically ventilate first and then isolate when needed or able. When we are searching from the interior our assignment and priority does not change. But we are typically entering through a door that causes ventilation, at least momentarily unless we are closing the door behind us. When we move throughout the structure and in and out of rooms, we isolate the rooms that we can and then are able to vent the room when needed. Once the fire is knocked down, we will be able to break out windows without isolating areas and when able, go back and open up the rooms that we isolated to allow for an improved airflow and environment.

WE ARE ABLE TO VENT WHEN:

- Fire is being overwhelmed or knocked down

- We are able to isolate

RULE 4:
LOCATE THE CIVILIANS

We target areas for search that are deemed best for civilian location and reduced time for the best outcome. If we enter a window, we will choose direct access to the room, area or type of room that we are prioritizing for search. When we move throughout the structure from the door, we target the same areas as we would by window but travel to them by different means as determined by the conditions.

RULE 5:
SEARCH ALL SEARCHABLE SPACE

When we enter for search, we are to search in the favor of our unprotected civilians by searching all searchable space in as little time as possible.

Photo by: Clackamas Fire

When we enter through a door, the whole assignment enters and moves throughout the structure to cover as much square footage as efficiently as possible and to continue throughout the structure until the assignment is completed. The rules are the same for entering windows, we shall remain disciplined to our assignment, enter with the whole assignment and continue to search all searchable space within the structure until the assignment is complete. This means that the first firefighter in the window will prioritize the threshold, occupy the hallway, search the hallway and rooms to their comfort and isolate when needed, while the second firefighter enters the window and searches the room. The two firefighters will meet up in the hallway and search methodically throughout the structure and possibly exit from another window or door. To leave an area that was searchable due to a "rule" from an acronym ("ves") with an unknown origin is negligible.

RULE 6:
SEARCH TO OUR COMFORT

We will spread out for search as our crew is more experienced, trained and conditions improve. The knowledge of fire attacks progress allows us to operate more effectively. When we hear that there is "water on the fire", "fire is knocked down", or alike, this allows us to spread out and continue throughout the structure more freely, this pertains to any and all search crews no matter the way of entry. When the conditions deteriorate, fire attack loses water or is being overwhelmed, our experience is novice or the structure is confusing, we will search closer to one another for comfort, yet still spread out enough that we are searching our own areas and not following one another path.

WHAT ALLOWS US TO SPREAD OUT DURING OUR SEARCH:

- Water on the fire, fire knocked down
- Experience of the crew
- Visibility
- Distance from the fire
- Knowledge of fire location
- We are able to isolate areas
- Fire is confined
- Knowledge of the building layout

RULE 7:
SEARCH COMMUNICATION WITHIN OUR CREW

Crews that perform orderly searches that are efficient and effective, emphasize most of their communication preparing for the assignment and keep speech minimal while performing on the fireground. The less we need to talk during search, the less confusion from muffled speech, the less breaths we will take, the less we will stop sweeping and moving and the more efficient we will be. Training on sizing up structures for search and working within a plan of how we will move throughout a structure will reduce our want/need to talk.

ESSENTIAL COMMUNICATION WITHIN OUR CREWS:

- Fire location
- Stair location
- "Victim! Victim! Victim!"
- "You go bravo" (This defaults that I search delta, could be a large room, rooms down a hallway or from a second-floor landing)
- Fire Attack lost water or being overwhelmed

NON-ESSENTIAL COMMUNICATION WITHIN OUR CREWS:

- "I have a bed"
- "Wall 1"
- "Here" "Left/Right", either "I'm over here", "Search over there", "To the right/left", etc is not a location and unless they can see you can be confusing and slowing our search

---

RULE 8:
SEARCH ON OUR WAY OUT

Typically, when we have a multi-prong approach to search, we meet up with another search crew somewhere in the building. We take advantage of our location within the structure and swap positions with the other crew, searching on our way out of the structure. The conditions will typically be improved, we are opening interior doors if the fire has been knocked down, but we aren't moving in and out of every room. We are quickly searching mostly with our eyes on the way out of the structure.

Using the Rules of engagement for search, allows us to approach the assignment of search with direction and purpose. This allows our crews to search through windows or doors with solid practices that directly benefit the outcome of our unprotected civilians.

SEARCHING WITHOUT A HOSELINE

ROB FISHER

Photo by: Jillian Stewart - iShootFire

If you have been on the job for any length of time, you should not be surprised to know roughly 2,900 civilians (averaged over the last decade) die each in the United States. In 2022, nearly 75% of these fatalities occurred in single and multi-family dwellings, as reported by the National Fire Protection Association (NFPA).[7] Firefighter Rescue Survey has crunched the data even further to show victims are located in the bedrooms and hallways 54% of the time.[8] Our ability to quickly access and search these areas will have a huge impact on victim survivability and continuing to decrease civilian fire fatalities in the United States in the years to come.

REASON

In recent years, the American fire service has seen a considerable change in operations surrounding search. In the last decade, the American fire service has experienced a significant increase of conferences across the nation providing focused training on search, vent-enter-search (VES), and

operating on the interior of a dwelling without the aid of a hoseline.

What was once viewed as an urban problem requiring a tactic only urban departments could perform requiring resources, now has bled into the suburban setting. The tactic itself was honed by urban city firefighters during the "war years." Today, much of the population has left the urban cities for more space and a suburban lifestyle. As a result, there are more dwellings occupied by people living outside the city. The fire problem is shifting to the suburban areas; not fully, but to a significant degree.

Searching without a hoseline or ahead of fire attack should not be used at all building fires. The tactic is built around being orientated to your surroundings on the interior. It is a tactic used most often in dwellings (single and multi-family) where there is a high likelihood to locate victims, and the space covered by a crew/company can be somewhat determined given a basic understanding of building construction and the common layouts within a region/jurisdiction, e.g., split-level, craftsman, colonial, cape-cod. Searching without a hoseline or search rope in a large open area like a big box grocery store is dangerous.

The question…Why would a company or a crew search ahead or without a hose line?

Simply...speed and agility. A crew/company can cover a greater area in a shorter time without a hoseline. The hoseline and other equipment slows down the progress of the search. Military doctrine would call this friction. The ability for a crew/company to search a dwelling quicker has proven to save more lives.

In his research project for his master's degree titled, *2021 Fireground Civilian Rescue Research Project,* by Chief Brian Brush, his research over 90 days in 2021 showed 10 cities with 5 or more separate fire incidents with rescues by firefighters reported in the news or department press releases. 89% of the total victims rescued were found pre-knockdown with a survival rate of 77%.[9]

Getting to victims quickly and being able to remove them is the critical piece to their survival. Our fireground tactics and how we get to them will have a measurable impact. Chief (Ret.) David Rhodes said it clearly in his recent editorial, "For departments to be good at search, they must have mastery of building construction, fire dynamics, and human behavior. Understanding—not just learning—in these areas allows the operator at the scene to make real-time decisions while implementing the highest level of risk management decision making. Totally removing options like searching above a fire, searching without a hoseline, and window-initiated searches indicates that your organization lacks mastery in those areas and only functions at the very basic recruit level in search technique.[10]

BUILDING CONSTRUCTION

Understanding building construction and common dwelling layouts are critical to perform a search ahead of a hoseline. Being good at dwelling layouts can be learned by driving around your company's first-due area. It can also be practiced on the EMS runs most companies respond to each shift (see training below).

Building construction will differ across the nation depending on where you work. Some areas (urban and suburban) have conventional construction mixed in with some modern engineered construction. When I was on the job, the dwellings in my jurisdiction were mostly engineered with only some conventionally constructed dwellings. Either way, a firefighter must know the environment they are working in to understand the impact fire dynamics will have on the structure.

Unless a dwelling is in demolition or poorly maintained, the walls should be able to withstand significant fire conditions long enough for most fire departments with good staffing to arrive and begin making a difference on progressively deteriorating conditions. This time can be further prolonged if a hoseline is quickly put into operation to begin confinement and to create and maintain searchable space for the search crew/company.

Interior doors can vary from solid panel to hollow core. Older dwellings are more likely to have solid panel doors simply because it was standard back in those days. You can also find them in modern construction in more expensive custom homes. However, more recently, the door commonly found is hollow core. It is also likely you would find this door in a legacy home that was remodeled in the last couple of decades, as solid panel doors can be cost prohibitive for many homeowners. Even a hollow core door can withstand a good amount of fire behind it. There are many examples posted on the Internet showing how well a hollow core door performed and prevented fire from communicating into a bedroom. These doors, and all doors for that matter, will last even longer once closed and water is applied to the opposite side of where the fire is impinging on it. This is where a search crew/company with a pressurized water can is effective at holding a fire until a hoseline can be stretched to the location.

A closed bedroom door can slow the spread of fire, reduce toxic smoke levels, improve oxygen levels and decrease temperatures dramatically for victim survival. The construction components within the dwelling are designed to aid in life safety. As firefighters, we must also use the construction features of the structure to aid us in preventing fire spread and as our protection as we move through the interior.

In experiments done by the Fire Safety Research Institute (FSRI) for the report, *Impact of Fire Attack Utilizing Interior and Exterior Streams on Firefighter Safety and Occupant Survival: Full Scale Experiments*, a victim behind a closed bedroom door below 3 ft (lying on the floor) was in a survivable space. "A victim located in a bedroom during a search with a closed door between them and the fire has a much higher likelihood of survivability than a victim with an open bedroom door. In every experiment, a victim in the closed bedroom would survive through the length of the experiment, including simulated fire department operations.[11]

EQUIPMENT

As an avid traveler, I always travel light with my carry-on and small travel backpack for speed and efficiency. This allows me the ability to move with limited friction through the terminal and on ground transportation much faster than those with much larger checked bags. Searching inside of a dwelling is very similar. The more you carry, the more tasks

Photo by: Eric Wheaton, Can Confidence

you need to perform, the more friction you create with your speed and efficiency.

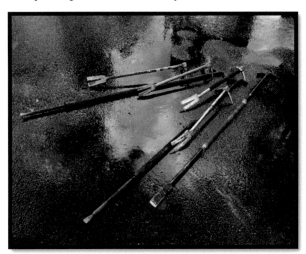
Photo by: Jillian Stewart - iShootFire

Searching without a hoseline helps to decrease your friction inside a dwelling. In place of a hoseline, the search crew/company should be carrying with them a pressurized water can (CAN) for protection. This is not to replace a hoseline. The CAN should only be used to push back a fire isolated to a small room. After a defensive search is completed in the room – if possible – the door is shut to eliminate the flow path and spread of fire into the hallway. The CAN is then used to keep the backside of the door wet until a hoseline can be stretched to the location. Wetting the back of the door prolongs complete door failure, even on the hollow core doors found today.

Next to the CAN, the Thermal Imaging Camera (TIC) is probably the next most important tool carried by a search crew. Many of the TICs today allow for good thermal readings of the environment they occupy. This is invaluable information when searching without a hoseline. A search crew can quickly determine if a hallway is tenable for search operations given our PPE and respiratory protection. A firefighter well-trained in TIC operations can easily discern if a hallway or large room (living room, kitchen, etc.) is close to flashover, requiring a hoseline before entering. Moreover, the TIC can assist us in orientation within a dwelling or building to help locate bedrooms, hallways, and stairways where we will likely find victims. It should be noted, the TIC is not an effective searching tool for the victim. It is best used to identify thermal changes and to confirm orientation or layout within a structure.

The Halligan bar is an important tool on the fireground. It is one of the most common tools carried by firefighters. It can be used for more than forcible entry. Depending on the demographics of your jurisdiction, the Halligan bar may be needed for forcible entry on interior doors. Most single-family dwellings will not have a forcible entry issue inside the dwelling; however, multi-family dwellings (specifically, center hallway apartments) will likely require additional forcible entry beyond the main entrance. In this case, a full set of irons will be needed or the use of two Halligan bars together (the

Photo by: Jillian Stewart - iShootFire

2nd used as a striking tool) can accomplish the job. Even if the Halligan is not needed inside for additional forcible entry, the bar should be used to hook doors for closing them in well-involved rooms. The Halligan is also a good tool for venting windows from the interior when a room can be isolated, eliminating any risk of a flow-path.

A better tool for hooking interior doors would be some type of all-purpose, multi-purpose, or officer tool. If a hook is desired by the search crew/company, it should be short and lightweight. This tool can be staged on the floor or in the area a search crew/company is searching to reduce having to move it through small areas. If needed, the crew/company can retrieve it from where it was staged.

THE OPERATION

It is often mentioned, Go/No Go starts with size-up. While I understand why size-up is critical in the decision-making process, I would argue it really starts with your level of training. Moreover, the level of training in your department and/or the department(s) you respond with. The idea of searching ahead of the hoseline when your department training does not include it as regular training or under Live Fire conditions is a potential recipe for disaster. The individuals performing a search without or ahead of a hoseline must have good skills, be in good physical condition, and understand their mission.

Size-up starts before the call. It starts when you are assigned to the station. Often, each neighborhood and residential dwelling is specific to a region of the county, e.g., Cape Cod, Ranch, Victorian, split-level, garden apartments. Legacy and modern construction are often built in planned communities where there is a similar layout, or a few used throughout the neighborhood or planned community. Likely, the dwellings you visit regularly on EMS or other incidents are like those in the neighborhood. You should be comfortable with the common layouts of the dwellings within our jurisdiction.

As your company starts to make the block or neighborhood, you should be looking at the era and materials used for construction in the neighborhood. Is it a modern or legacy neighborhood? If it is more legacy, has there been recent remodels in the area that will blend the two types of construction?

On arrival you should be sizing up the dwelling from your rig. How many floors is the dwelling? Where is the fire currently at and where does it appear to be going? Where are the bedrooms located? If it is a two-story dwelling, where are the stairs likely located?

Next, you should be evaluating the volume, velocity, density, and color of the smoke. The highest velocity often indicates being closer to the seat of the fire. Velocity is created by excessive heat and energy, of which are both present near the seat of the fire. Good volume of smoke with no active fire can indicate the fire is deep seated and a ventilation limited fire condition. Visible fire is a good thing because the fire you see is energy being released. Your biggest problem is radiant heat, which can be limited through shielding. You must remember that smoke is fuel. Being able to anticipate or predict changing condition as a result of the fire's progress and/or our actions is critical to being able to work without a hoseline in a dwelling.

Photo by: Jillian Stewart - iShootFire

As you approach the dwelling, you are looking to better identify bedrooms and some sense of the layout. If you are entering without a hoseline, it would be best to make your initial entry through the front door because it puts you in the "ways" as in stairways and hallways. Moreover, it puts you at the foyer or main entrance where victims are often found, and you are best able to orientate yourself to the layout of the dwelling. If the engine has a hoseline through the front door upon your arrival, consideration should be given to a window-based search directed at one of the bedroom windows. However, if the conditions on the interior support basing your initial search operations from the interior, it may be quicker and easier to access the bedroom from the hallway(s).

In departments with good staffing and plenty of resources, the truck(s) on the fire will most often perform the search. In most cases this will be performed without the aid of a hoseline. In many departments, however, this is not their reality. Given that 80% of the ladder trucks manufactured these days are quints, the suburban department has a more challenging decision to make regarding search vs. stretching a hoseline. This is a discussion for another time, though. Either way, you and a partner might find yourselves ahead of the hoseline if the engine is delayed with the stretch or if you arrive ahead of them.

In my jurisdiction, with fire-based EMS, often the BLS or ALS unit in our firehouse were assigned search on their arrival. These units were outfitted with a CAN, TIC, set of irons, and a 5 ft. inside hook. They were self-sufficient for a primary search. Their search was supported by either a hoseline stretched off the quint or an engine arriving just after their arrival.

The job of the first hoseline stretch is to support the search. In most cases, it is used to quickly confine the fire to the room(s) of origin. Thereby, reducing the spread of fire and superheated toxic gases. Some departments have sufficient staffing to stretch a backup hoseline in addition to the other tasks on the fireground. The backup line is tasked with protecting the search and the hoseline assigned to fire attack. The location of the backup

$$Ot = \frac{(T_c + Sq_f)F_g}{E_T + R}$$

line should be near the hallway on a one-story single-family dwelling or near the stairs on a two-story dwelling if the search is being conducted above the fire. This places the backup line in the most opportunistic position to engage should they be needed.

The search crew/company entering the structure should perform a life, fire, layout assessment upon entry through the front door to determine, if there is life in the immediate area of the entrance? Where is the fire located within the dwelling? And what is the layout of the dwelling; does it match what you saw/thought upon your arrival?

While you are doing your life, fire, layout...yell out, "Fire Department...is there anyone here?" Listen intently for a victim(s) to callout. A victim yelling for help happens more frequently than one would think. If you have not already masked up, close the door and do so. As you are masking up, observe the fireground and the other operations around you. Is the engine stretching a hoseline to your location? If so, let them know where the fire is located and where you will be starting your search. If there is no engine or hoseline yet, you must begin your search for fire before beginning a primary search. The fire MUST be controlled or contained before you can continue with a primary search.

As you re-enter the dwelling, use your TIC to help locate the fire. Be sure to close the front door to prevent a flow-path behind you. One of the tactics of a crew/company searching inside a structure is to perform anti-ventilation actions...eliminating air getting to the fire. You and your partner need to hustle to the fire's location and quickly work to isolate it to the room of origin. The CAN may be needed to push back fire, so a defensive search can be completed before shutting the door to the room. Then the CAN should be used to wet the back of the door to help prolong its failure. While all this is going on, the other member should perform a search in the area outside of the fire room for victims. Once you know where the fire is located, this needs to be communicated to command – if one has been established – or the engine arriving to perform the stretch. As soon as the hoseline is stretched to your location, you and your partner can peel off to complete the primary search on the rest of the dwelling.

You can only move through the interior where conditions allow. The intent of searching without or ahead of a hoseline is not to go diving into a super-heated environment at pre-flashover. You must be smart about your movements between spaces (boxes) within the dwelling. Like working a hoseline through the building, you need to occupy one space (box) at a time before you move on to another. You should use the walls and corners for shielding from radiant heat from the fire. Staying low in a tripod position will reduce your profile from convective gases passing overhead. Using your TIC to periodically scan your travel path will help you identify deteriorating conditions.

It was mentioned above…door control is the key to reducing flow-path and preventing fire spread. It is also known as anti-ventilation in some of the older tactics manuals. Anti-ventilating throughout the interior provides the search crew/company safety outlet should things begin to deteriorate. Along with anti-ventilating, the search crew/company should be noting where bedrooms are when searching a floor above the fire. These rooms can be used to seek refuge and safety isolate should conditions deteriorate as well. Should this happen, stay calm and radio command with priority traffic of the changing situation and your location. The backup line – if staffed – will be moved to resolve extending fire into the area you and your partner are searching.

Ventilating as you search. This can be done by closing the door to the room you are searching to prevent creating a flow-path. Terms like, "water on the fire" help searchers know the incident is being stabilized and aggressive ventilation is now possible. Given the #1 cause of civilian fatalities in residential dwellings is smoke inhalation, rapid ventilation of the structure helps both the civilian and firefighter by removing the toxic smoke that is killing civilians and obscuring the visibility of the searchers. Consideration should be given to removing the entire window with a tool when using it for horizontal ventilation. You are only using a portion of the total window area when it is manually opened. This may be effective with a light-smoke condition on a small fire quickly contained by fire attack. On much bigger incidents, taking the window would be appropriate. Again, this should only be done if the room can be isolated from the rest of the structure, e.g., bedrooms and offices. Additionally, any window screens should be pushed out when using the window for ventilation. The screen will reduce the effectiveness of the ventilation. It will eventually become fully occluded by the sticky, particulate smoke making your ventilation opening ineffective.

Searching without or ahead of a hoseline is not just an urban fire department tactic, it can be done by many suburban departments as well. What urban departments have over smaller departments is resources and experience. This allows them to perform simultaneous operations with a higher operational tempo. A department's operational tempo (Ot) is result of the type of construction and size of the structure on fire divided by the experience, training and resources. The Ot of a department can be increased by bettering in their experience, training and resources. The two easiest to improve are experience and training, as resources require hiring or additional companies being added to fireground incidents…time plays a factor in their effectiveness.

TRAINING

In his September 2024 editorial titled, *The Lack of Search Training is Appalling*, Chief (Ret.) David Rhodes was critical with departments fortunate enough to have resources on the fireground. "In many departments, search is something that happens after extinguishment. This is sometimes necessary depending on the circumstances and staffing levels but, unfortunately, is the practice in many departments that have the resources and staffing to conduct simultaneous tasks but lack the necessary training."

As we did in our adolescence move through basic addition and subtraction to more complex mathematic equations, we must as a fire service progress beyond basic fireground operations. Our ability to quickly process the fireground and adapt to it should improve with experience and training. Ten years on the job should mean ten years of improving your craft of the job. If we were to compare a professional athlete to a high school athlete, the game has slowed down for the professional player. While they are both playing the same game with essentially similar rules, the professional athlete has more experience and training than the high school athlete to quickly adapt; thereby, slowing down the game. The fireground is no different. It will slow down with experience and training.

After recruit training and a period of probationary time, a firefighter must seek out more advanced training and/or conferences if their department does not provide this level of training. At the time this article was written, here are four cadres or programs you should seriously consider taking to improve your experience and training on the fireground.

> **Brothers In Battle:**
> 1. 1st Due: Search 4 THEM
> 2. VES: Beyond the Door
>
> **FDTN:**
> 1. Truck Operations
> 2. Rescue Operations
> 3. Fire Combat
>
> **Insight Fire Training:**
> 1. Introduction to Tactical Thermology
> 2. 8-Hour Live Fire Training
>
> **VentEnterSearch:**
> 1. CAN Confidence
> 2. The Search

What these four cadres provide is realistic training on how to occupy the interior of a dwelling without a hoseline focused on the mission of search and saving lives. The key is realistic training. To understand your limitations, you must train in similar environments you will be working. Theater smoke does not provide this level of training. This is fine for initial training and frequent drills. But to develop your experience, you must be able to evaluate convective heat movement in a limited visibility environment.

There are several other cadres/programs that provide similar training. The four mentioned above are just the ones I know most of the instructors personally. They are some of the best in our profession.

Use your EMS runs as an orientation exercise on the layout of the dwelling you are responding to the medical call in. On arrival to the EMS incident call out the location of the kitchen and the bedrooms. If it is a two-story dwelling, add the location of the stairs leading to the 2nd floor and the main hallway on the first floor. Then, following the EMS call debrief as a company to see how you did.

This is just a simple way to practice your layouts within a dwelling.

CLOSING

Why do we respond to fires?

In his editor's opinion back in April 2023, Chief (Ret.) David Rhodes addresses common sense operations on the fireground and our true mission…to search, rescue, and save lives. Rhodes wrote, "The data is in and confirms what I was taught in recruit school 38 years ago: TIME is against those trapped. Smoke and toxic gases were identified as the cause of death in fire victims more than 50 years ago, which has been reconfirmed by recent studies. Data now tells us that the victim's chances of survival are 79% if removed within 2 minutes of arrival, 73% in 2-4 minutes, and 65% in 4-6 minutes. Again, the fire is not killing them; they are drowning in toxic gases, and the only thing that makes it better for them is to get them out.[12]

It takes speed, agility, experience and training to search without or ahead of a hoseline. Ben Shultz, lieutenant with West Palm Beach Fire, has a lecture titled, *Every Second Counts*. In his presentation he compares the benefits of speed and agility of an alpinist with a mountain expedition like Everest. Each has their benefits. Specific circumstances afford the alpinist more benefits at the risk of exposure. But it is the circumstance and benefit the alpinist exploits to be successful climbing. On the fireground, we need to consider the benefits of reducing our friction, understanding the risks associated with doing so, to increase our speed and agility to save more lives.

Searching ahead of the line or without a hoseline is not a new tactic. It has been done since people occupied an enclosed space and built fire. Techniques used for this tactic are nothing new either. James Braidwood covered some of them in his 1866 book, *Fire Prevention and Fire Extinction*. What has changed, however, is the building, the contents inside and human behavior making for a more challenging environment to work. This does not relieve us of our duty to protect and save lives, though. It just means we must be smarter and better prepared to handle a rapidly changing fireground.

[7] Shelby Hall, "Fire Loss in the United States," National Fire Protection Association, accessed September 16, 2024.
https://www.nfpa.org/education-and-research/research/nfpa-research/fire-statistical-reports/fire-loss-in-the-united-states

[8] Nick Ledin, "The First 3000 Rescues," Firefighter Rescue Survey, accessed September 16, 2024.
https://www.firefighterrescuesurvey.com/uploads/2/5/3/6/25366566/027_victim_location_room_v_total_recorded_rescues.png

[9] Brian Brush, "2021 Fireground Civilian Rescue Research Project," Firefighter Rescue Survey, accessed October 9, 2024.

[10] David Rhodes, "The Lack of Search Training Is Appalling," *Fire Engineering*, September 2024, 6.

[11] Robin Zevotek, et al. "Impact of Fire Attack Utilizing Interior and Exterior Streams on Firefighter Safety and Occupant Survival: Full Scale Experiments," Fire Safety Research Institute, (2018): 182, accessed September 16, 2024

[12] David Rhodes, "We Go to Fire to Search," *Fire Engineering*, April 2023, 8.

SEARCHABLE SPACE: THE SIZE-UP
SEAN DUFFY

Photo by: Csmeadphotography

WHAT IS OUR JOB

When occupants become trapped in a fire, our number one priority is rescue, but just because we arrive does not mean they will survive. However, the way we choose to access and remove them can make a difference. Our job on the fireground is not to decide who lives and who dies but rather to give the best survival odds to those trapped by fire that we possibly can. Simply put, we are not there to guess their odds of surviving, we are there to give them the odds. The search 360 is one of the best ways we can gather information quickly in regard to making decisions on where we will dedicate our initial crews.

There has been some confusion amongst our profession with using terminology such as tenable space or survivable spaces (survivability profiling). To make things simple in an already chaotic situation, using the term searchable space will more accurately depict what we see during our size-ups.

SEARCHABLE SPACE

Searchable Space is defined as any space that we can occupy and operate in with full protective clothing. If we can make the determination that such a space is what exists, we MUST dedicate crews into it for search and rescue, as civilian lives depend on us to decrease the time it takes to locate and remove them. There are a lot of things that we cannot control but our commitment to putting our citizens first and being prepared is one that we can! We shall not rob our citizens of any chance for survival by not keeping search a priority, this includes training our minds to look for opportunities.

Photo by: Csmeadphotography

Remember it is human nature for our eyes to track movement and light so we must not become too consumed by the fire itself and look at the scene through a much bigger lens. Knowledge of smoke reading, fire behavior and building construction/ layout is paramount but we cannot lose sight of the spaces we have available to search in the process.

TWO-OPTIONS

When it comes to searchable spaces we have two options, we can either create and maintain spaces or we keep from losing the spaces we already have. In both cases fire attack and isolation play a big role. Our objective when committing to a search is to concentrate our efforts on where we will have the best chance of finding victims. If searchable spaces exist on arrival, dedicating crews to the interior prior to fire attack could be of greater benefit to our civilians. The reason this decision can be of great benefit is because although searchable spaces may exist at the time of arrival, the odds of survival decrease as the amount of exposure to toxic gasses and heat increases.

It is important that we comprehend what our spaces are telling us. Not just for our search efforts but also what it means for potential victims within that space. It has been shown that occupants who may have become incapacitated may still be able to survive their exposures if rapidly located, removed and taken to appropriate medical care. The earlier into an incident that our victims are removed from the structure, the less toxic exposure they will receive, ultimately increasing their chances for survival.

STAY FOCUSED

When analyzing spaces for search, stick to the objective and be sure to gather all the information and facts to formulate your strategy, then apply resources to actions that will do the greatest good at that moment. Be intelligently aggressive but make your decisions based on a solid foundation of core knowledge. During the analysis of searchable spaces be sure to take into consideration where the fire has been, where it currently is at, and where we project it to go (BAG the fire).

> **BAG the FIRE**
> Where has the fire **BEEN**
> Where is the fire **AT**
> Where is the fire **GOING**

While observable conditions from the exterior are important to consider, they cannot always tell us everything that is happening on the interior. Searchable spaces exist inside of structures that we cannot fully see from an extremely limited exterior view. When we stand outside of a structure we have a 2D view of the situation as we can only

SECTION 3: DECISIONS

and other tactical priorities. Ultimately the choice to conduct search ops, and the specific tactics deployed in support of that decision, are determined by department policy, the IC, and the specific incident. However, we must start believing and demonstrating to our citizens that we can and WILL make a difference every day, otherwise, who will?

Photo by: Csmeadphotography

see the 4 sides making up the building. Once we make entry via door or window, we now gain a 3D perspective as the structure starts to be divided up by walls, furniture and rooms. Closed doors are a prime example of an interior sizeup for searchable space. Sure, we may have some indication that a door could be closed from the exterior, but the only way to truly know is to enter that space and confirm it. This is a critical part when it comes to searchable space because closed doors are open opportunities that compartmentalize our victims from the fire itself. A closed door decreases temperatures by nearly 900 degrees, improves oxygen concentrations to 18% and reduces toxic gas levels to 100ppm. Despite floor to ceiling combustion on the other side of the door, a victim can still survive in this space if identified and searched quickly.

Once a space is identified upon a 360, it should be relayed to command and in coming units that you have searchable space and what the location is. This will help set the stage for the search effort

THE FIRST-IN INTERVIEW
SCOTT CORRIGAN

Photo by: Jillian Stewart - IShootFire

THE NECESSARY MINDSET

A dwelling fire with entrapment is the foundation upon which we build the mindset of the Fire Officer. It is the foundation that provides the balance for conversations before and after the event. It is the context by which we frame our training scenarios, create our task timelines, and guide our raw tactical decision- making. The dwelling fire with entrapment is the "The Event" that we prepare for; it is our Arena. In our Arena, we must determine the location of the occupants and main body of fire.

To help further clarify, I will provide the written context that exists in the Tactical Documents at my agency: *"Residential houses are assumed occupied until a primary search has been completed by the fire department."* These aren't just words on paper; they are engrained in our culture and operational expectations. It is a finite mindset focused on a successful rescue when our citizens find themselves imperiled by fire. Our mission is to find them as fast as humanly possible and escort them to medical providers.

WHAT IF VERSUS WHAT IS

We have to acknowledge the two mindsets that we find ourselves in, the infinite and the finite. The infinite mindset is the one that we use to remain limitless in our creativity before the event. We identify the variables and categorize them as those we can control and those we cannot. Our goal is to arrive at a dwelling fire with entrapment having an infinite mind, operating within a finite standard operating procedure. *Before* the event is the time to exercise and dialogue on the infinite "what if". At the event we operate based on the finite "what is."

Our interaction with citizens at a dwelling fire needs a finite framework. If your agency leaves how to interact with citizens at a fire to "player's choice," you will get information that feeds back into your decision-making that is not standard. It will be hard to build a standard response to non-standard input. The key to the framework is setting goals based on priorities. The goal is to *find* and *remove* trapped citizens. The priorities are set by RECEO.

FOCUSING ON WHATS IMPORTANT

To help guide the conversation at your agency, I submit the following questions:

> **What are the goals when interviewing a citizen at house fire?**
>
> **What information are you attempting to retrieve or disseminate?**

Once these things are defined organizationally, your first-due officers need a simple framework by which to identify actionable/*high-value* information. This is the target information that leads to operational expectation or change needs to be communicated to the first alarm units. The targeting mechanism in this framework also provides a buffer from potential distraction of first-alarm units by filtering out low/lower-value information that has little or no bearing on initial actions.

Photo by: Jillian Stewart - IShootFire

The information that we are targeting, the *high-value* information that triggers operational change and/or is communicated to all first alarm units, is:

- ***Someone is inside***
- ***The exact location of the person***
- ***The location of the fire***

In order to obtain this target information in a succinct and effective manner, the very first filter I use is to ask the citizen if they live in there. This immediately clues me in as to the intimacy with which the person is able to provide me information.

SECTION 3: DECISIONS

A resident of the dwelling on fire will have much more intimate and reliable information as to what is happening inside the house than the neighbor.

1) Do you live here?

If it is indeed someone who lives in the house, the next question I ask is, "Is everybody out?" listening for the high-value answer of, "No." and filtering out any answer suggesting that the dwelling is unoccupied. Again, the high-value answer is the one that is actionable and communicated. Being told that everyone is out cannot have any impact on our operations. If they state that someone *is* inside, with or without a location, it is high value information and should be transmitted on the radio. When I receive a response that indicates people are still inside, I ask for the location. I ask, "Where are they exactly?" which allows me to give specific directions in deploying the members for search, as well as allowing me to orient myself to their location as I enter the building.

2) Is everyone out?

3) Where are they exactly?

The next question seeks clarification on the body of fire. I am shocked at how specific people can be if you ask. The question you ask the citizen is, "Where is the fire?" The answer you get back is either vague or specific. The vague has no value, while the specific holds *some* value, although neither should be transmitted on the radio. The information is held and used by the Officer during the fireground size up and the 360, if done.

4) Where is the fire?

When completed in a manner that is succinct, purposeful, and targeted toward immediately identifying high-value information, this interview at a house fire can further sharpen my focus and tactical decision-making and clarify the location of assignments, without slowing the speed of deployment of the first alarm or negating the absolute truth that our mission, as stated above, is to successfully rescue our citizens by finding them as fast as humanly possible and escorting them to medical providers.

ASKING THE RIGHT QUESTIONS

DANE YAW

Photo by: Jillian Stewart - IShootFire

When arriving at a chaotic fire scene, it can be difficult to know where to begin. In the midst of blaring sirens, thick smoke, and frantic bystanders, the ability to cut through the noise and focus on what matters most becomes critical. Over the years, I've had the privilege of learning from some of the most seasoned firefighters who are now Chiefs, whose calm under pressure and ability to ask the right questions have shaped my own approach to search and rescue operations. In this article, I aim to convey the invaluable lessons they've shared, emphasizing how effective questioning of bystanders can be a powerful tool in guiding our efforts and saving lives.

We have all seen firsthand how quickly fire scenes can turn chaotic. During these high-stress moments, every second counts, and often the most crucial information comes from the bystanders—the people who saw it happen, who might know who's inside, or even how the fire started. However, in the rush of responding to an emergency, asking the right questions is not always easy. It's not just about asking questions but asking the *right* ones in the right way. Our goal is to get accurate, useful information to guide our search and rescue efforts while filtering out the distractions or misinformation that can slow us down. I'll share insights on the importance of bystander questioning, based on years of experience and learning from several seasoned chiefs in the field.

INITIAL QUESTIONS: THE SEARCH FOR LIFE

One of the first questions we need to ask on arrival is:

"Does anyone live here? Is anyone inside?"

This might seem obvious, but how you ask it is critical. One of my informal mentors and now Deputy Chief of Operations with Oklahoma City Fire, Chief Mike Walker offered some sage advice. "Make them show you the exact window if possible", bystanders can often be vague, especially in a panic. They might answer with something like, "I live across the street," which doesn't give us any useful information. Instead, we need to push for specific, clear answers. For example, ask them to point to the exact room or window where they believe someone might be trapped and in need of rescue. This focused approach helps us prioritize our efforts and zone in on the areas where life-saving rescues are most likely needed. This information will also point us in the right direction in determining our tactics. Is this going to be a conventional search or a VEIS opportunity will be based on the information gleaned from questions and size up of the structure.

PETS AND PRIORITIZATION

In the confusion of a fire, people often forget about their pets. That's why an important question of clarification is, "Are there any pets inside?" It's common for neighbors to know if pets live in a home, even if the residents are too panicked to mention them. But the critical part here is to clarify what the person means when they say things like, "My baby is inside!" People often refer to their pets as children, which can lead to misunderstandings and a potentially flawed search prioritization. It's important to ask specifically what type of baby they're talking about to ensure that limited rescue resources are properly allocated.

CUTTING THROUGH THE NOISE

Another challenge we face is getting sidetracked by long-winded or irrelevant stories. We've all asked...

"Do you know if anyone is inside?"

...only to get a rambling, unclear answer. In moments like this, it's crucial to steer the conversation back to what matters. Chief Mike Paschal, one of OKCFDs Battalion Chiefs and considered by most within the department as a subject matter expert on fire ground command, bluntly explained:

Photo by: Jillian Stewart - IShootFire

> "Those that know the least always know it the loudest."
>
> ~ Mike Paschal

SECTION 3: DECISIONS

While it may seem harsh, it's important to politely but firmly shut down conversations that aren't helping. My current Battalion Chief and longtime friend, David Meyer has added to this topic:

> "Anyone that answers with more than short answers, I shut down quick. Time is our enemy, and the victim has no time for a speech."
> ~ David Meyer

This approach is about maintaining focus. Time is a luxury we don't have in an emergency, and we need to stay on task.

Once we've gathered key information, it's critical to keep the bystanders engaged. After asking about people or pets inside, I make sure to ask for their name and request that they stay in one spot. Chief Randy Cornelius, Oklahoma City Fire Department second most senior Battalion Chief shared with me:

> "This helps establish accountability and ensures that if we need to follow up, we know exactly who to talk to."
> ~ Randy Cornelius

This also reassures the bystander that their information is important and prevents them from wandering off in case we need further details. Sometimes, these individuals become key witnesses who can provide additional insights later when you begin your investigation.

FILTERING MISINFORMATION

Unfortunately, not all bystanders will have accurate information. Some may feel the need to be involved and share what they "think" happened, even if it's not reliable. That's why it's important to gauge a person's credibility. As Chief Meyer previously pointed out:

Photo by: Jillian Stewart - IShootFire

> "If they seem like they are just talking nonsense, I ditch them hard."
> ~ David Meyer

We need to be cautious about following leads from people who might be offering guesses rather than facts. Misleading information can waste valuable time and, in the worst cases, put rescuers in unnecessary danger.

UNDERSTAND THE CAUSE

Knowing how the fire started can be useful, even before the fire is fully under control. A question like, **"Do you know what happened?"**, can help firefighters anticipate what they might face when entering the building. For example, a fire that started in the kitchen could indicate there may be less structural damage compared to an electrical fire that has spread through the walls.

As Chief Cornelius explained, asking questions like "Was anyone cooking?" or "Have there been any electrical problems?" can provide clues that help firefighters assess the risk and prioritize their approach.

Summary of Key Questions to Ask:

1. **Does anyone live here? Is anyone inside?** Always ask this first, and get specific information, such as the exact locations or rooms where people might be.

2. **Are there any pets inside?** Neighbors or witnesses often know, even if the occupants forget to mention it in the panic.

3. **What happened?** If possible, get details on how the fire started. This can help with understanding the building's condition.

4. **What's your name? Can you stay here?** Get their name and ask them to stay put. You might need to follow up for more information.

5. **Filter out the noise.** Stay focused on direct, short answers. Those who ramble might not have the information you need.

Questioning bystanders may seem like a simple task, but it's a skill that takes practice. In the chaos of an emergency, the ability to gather quick, accurate information can be the difference between life and death. It's something that every firefighter must learn to master, and something that continues to evolve with each experience we face.

This approach highlights the importance of gathering and filtering the right information quickly and decisively, ensuring that every action taken is one step closer to saving lives.

CONCLUSION

As firefighters, our ability to save lives and protect property starts the moment we arrive on the scene. The right questions, asked in the right way, can give us the critical information we need to guide our actions. By staying calm, focused, and efficient, we can cut through the noise, prioritize our efforts, and make decisions that save lives.

COMMAND FOR THE RESCUE
BRIAN BRUSH

Anticipation is the most intelligent form of aggression and this an essential quality of a prepared incident commander.

EXPECTING FIRE

We have all heard the mantra "Expect Fire and Expect Victims", it will likely appear dozens of times in this book. While having a vigilant mindset for the "if" is important; knowing what to do, or better yet, having everything in place to execute the "then" is what changes outcomes.

In my experience, most fire departments and incident commanders respond to fires with an "Expect Fire" plan. Given a dispatch to an apartment fire of: BC1, BC2, E1, R1, T1 E4, E6 and M6. We would expect, either through best practices, NFPA 1710 recommendations or department procedure that the first arriving engine (E4) will initiate fire attack, next engine (E1) will secure a water supply for fire attack and support it with a back-up line, and finally, if the number of floors or type of occupancy calls for it, an exposure line will be stretched (E6). IF there was fire upon arrival THEN these

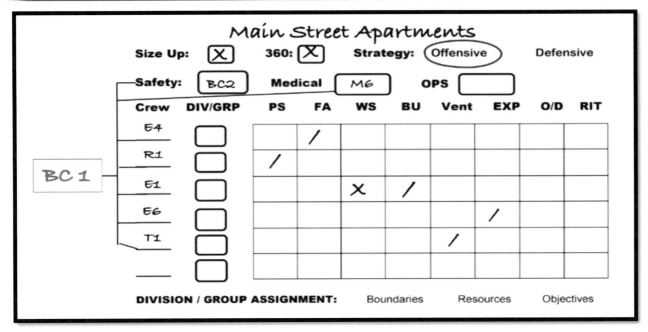

assignments would almost automatically roll out in this order as is reflected in the tactical worksheet below.

While organizing charts isn't the type of "aggressive firefighting" most of you came here for, staying ahead of an incident and issuing clear assignments proactively at the command level is what allows for it to occur at the tactical and task level. The tactical worksheet is the best way to both "show our work" and conduct an exercise that displays to command officers the impending challenges.

With it written down in black and white, we can see half of the initial resources to this alarm are dedicated to fire suppression, supply, or support. We can also show that if we convert this tactical worksheet to ICS organization chart from left to right, the incident commander is already at a span of control of 1:7. If you include the demand of speaking to dispatch the span of control is 1:8, and it will only grow with each additional unit arriving because by design, right now every resource is a direct report to command.

The key takeaway here is that almost all departments have some version of a fire attack plan that is well known from top to bottom, unfortunately for most, this single plan for a single problem pushes limits in both resources and attention and there is very little room for any additional challenges.

At the task level, fire attack is making a room or a hallway and opening the nozzle. At the tactical level, fire attack is the company officer directing and coordinating that nozzle firefighter and the back-up supporting the advance. At the strategic level, fire attack is the IC ensuring that once fire attack is initiated, it is supplied and supported with adequate resources. We can say that given this scenario the plan supports all these objectives at the task, tactical

Photo by: Jillian Stewart - iShootFire

SECTION 3: DECISIONS

and strategic levels but it also shows that this "Plan A" is only matching that demand.

It is 2:58 in the morning and you are responding to a reported house fire. As you are pulling out of the station the dispatcher advises you that she had the occupant on the line stating she was trapped but has lost contact. Two minutes later you arrive on scene with a working fire, two companies and you have confirmation of a victim trapped.

Consider what we just discussed; the entire first alarm rolled out the door with that expected fire plan. It will still work, but will it come at the expense of the rescue?

The most important thing to recognize at the command level is that the trapped victim in the fire building is an ADDITIONAL problem to be solved with its own set of tasks, resource, and communication demands. The proper terminology for this is an Incident Within an Incident (IWI), the rescue incident is happening within the fire incident.

To use a football analogy, the best time for a quarterback to audible out of a play call or make a shift in formation is that time when everyone is in position but before the ball is snapped when you are still at some level of control. Once the ball is snapped you are adapting in real time to dynamic conditions that have momentum for or against you.

EXPECTING VICTIMS

Earlier in the article we discussed that at the strategic/command level, fire attack is much more than just getting the nozzle to the fire room; it is ensuring all the bases are covered and supported, attack, back-up, supply and potentially exposure. Provided the "Main Street Apartments" example and the tactical worksheet above, these assignments took 3 companies, most of the initial alarm assignment. So, what about "expecting victims"?

A victim trapped in a fire building will need to be located (search), removed (rescue), triaged, possibly treated, and transported (medical) and, unless the victim was in the last accessible location, the search will need to be resumed (search). If we

responded to a collapse with a victim trapped in a house with no fire, it would be a stand-alone multi-company operation with an incident commander. Just like fire attack, if we are truly expecting victims, we must anticipate that this is an incident within an incident with a resource and communication demand beyond just assigning search and having an ambulance on the dispatch. Within 5 minutes of arrival Ladder 6 has located and is removing a victim, they are requesting EMS to the front door. As this radio traffic is coming across the incident commander is tasked meeting their new requests and recognizing that the assignment that Ladder 6 had (primary search) has not been completed and Ladder 6 as a resource is no longer available. The same goes for the ambulance assigned medical. Once medical receives a patient, they are working and no longer an "available" resource.

SECTION 3: DECISIONS

Once fire attack is initiated it is supported with a back-up line and a continuous supply. The same approach must be applied to rescue. Once search is initiated, it must be supported with an available medical, and backed up as soon as possible another available company, so that "IF" a victim is found "then" they are at the ready to assist with the removal or to continue the search from where that victim was located.

Photo by: Csmeadphotography

CALLING THE AUDIBLE

Here is where the parallel must be made and appropriate performance forged. Fire Attack and Search are the highest incident priorities, the order of performance will be dictated by conditions and resources.

"Although survivable spaces exist at the time of fire department arrival (Section 7.4, the survivability potential decreases as the time of exposure increases. When resources permit, interior search and rescue operations can and should proceed simultaneously regardless of the fire attack tactic selected." ULFSRI

Simultaneously initiating fire attack and search has been and remains the common practice in countless cities and response areas where a fire engine and a truck company either respond together or arrive to an address in close order, but it shouldn't matter what type of apparatus arrives, especially in a known rescue situation.

If you are in a department that is primarily fire attack focused with a deployment plan that has attack, support, supply or "level 1 staging" as pre-determined assignments for the first arriving rigs you need to practice a "Reported Victims Trapped" play. That "play" ensures the tasks of fire attack and search are both initiated as soon as possible with limited communication and delay and then supported as full Fire Attack and Rescue Operations as needed.

You are dispatched to reports of victims trapped in a house fire with this first alarm:

BC1	BC2	E1
R1	T1	E4
E6	M6	

As soon as you receive this information from dispatch you call the audible. "Battalion 1 to all units responding to Main Street. We will be operating as a confirmed rescue. Engine 4 (1st due) upon your arrival initiate fire attack, Engine 6 (2nd due) bypass the hydrant and initiate search, Engine 1 (3rd due) you will now pick up the hydrant. Battalion 2 upon your arrival you will assume rescue group supervisor."

I lead the article explaining that anticipation is the most intelligent form of aggression and an essential quality of a prepared incident commander, this audible is that action of anticipating the potential a scene like this has.

The common misconception is that you are "running the fire and you aren't on scene", this is not true. With one pro-active radio transmission you have adjusted a conventional plan that was set up to COMMAND 1 operation (fire attack) and you are now ready to COORDINATE 2 operations (a fire and a rescue). You also eliminated the "paralysis of analysis" for that first arriving company trying to decide attack or rescue and reduced follow up communication of who is catching water supply and where you want BC2 when they arrive.

Engine 4 goes on scene with "A working fire in a single-story single-family residence, fire out two windows on the alpha side".

When you arrive to take command, Engine 4 is stretching a line to the front door with Engine 6 following them with a set of irons to start the search. You send Rescue 1 to do a 360 and look for VES opportunities and you assign Truck 1 Vent. You can see Engine 1 making a hydrant connection down the block.

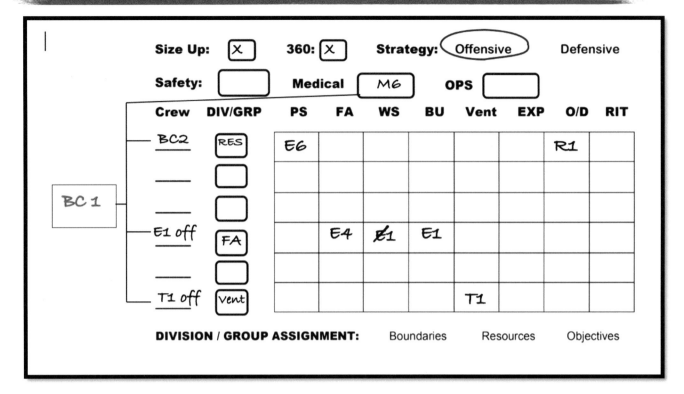

With your audible during the response, you pre-loaded the command structure and initial actions. Now with your arrival it is time to formally take command, clarify and confirm this plan to coordinate and support the operations.

"Battalion 1 is on scene I will be taking Main Street command, I've got Engine 4 with a line to the front door, Engine 6 on the primary search from the Alpha side and Truck 1 headed to the roof."

"Command to Engine 1 officer, when you secure that supply you will be Fire Attack Supervisor, you have E4 and your crew working for you with the objectives of fire control and protecting the rescue operation.

"Command to Battalion 2, when you arrive, I need you to the Alpha Side as the Rescue Group Supervisor, you will have Engine 6 and Rescue 1 working for you." This assignment is one of the most critical to make prior to their arrival. Battalion 2 will now arrive and assess the scene as the Rescue Group Supervisor. His size up and planning will be laser focused on the rescue from the start, and he will go directly to those resources already initiating tasks. This sets a concentration and discipline to a Plan A Rescue that is very difficult to achieve when you arrive to an incident thinking fire and have to shift to rescue.

Now when we look at this tactical worksheet, you can see that calling the command audible reduced the span of control from the IC to 7 units performing tasks, down to the IC to 4 supervisors running 4 tactical operations. Not only does this reduce the span of control early in the incident, it will likely stay this way as any additional units will be assigned directly to one of these group supervisors.

Moving division and group supervisors into place at the tactical level brings decision making closer to the problem, saving the time that is often lost to communication and clarification to a more remote IC. In the context of a rescue at a fire, having a fire attack group supervisor and a rescue group supervisor in place and position to work together in the front yard really tightens the coordination between the two operations. Communication between them likely being nearly 100% face to face significantly reduces the potential for misunderstanding or lost accountability.

TIME IS LIFE

I know that majority of the firefighters reading this are firefighters or company officers and you know you can "buy time" for our victims through

SECTION 3: DECISIONS

quickly masking up, drilling in operations and setting up the apparatus for success. When you move to Chief or Command Officer the position is different, but your ambition must be the same. You can "buy time" for our victims by quickly and proactively decentralizing command to reduce the delay of "mother may I" communications, you can drill with your company officers in division and group supervisor positions to empower them to be confident in their decision making putting more problem solvers closer to the problems and you can set up your command system for success. Stop wasting time, know that success on the fire ground and survival for our victims depends on alignment of tasks, tactics, AND command. Expect fire, expect victims and expect the IC to be in a position to coordinate the two. Anticipation is the most intelligent form of aggression and an essential quality of a prepared incident commander.

COMMUNICATION CONSIDERATIONS
NICK LEDIN

Photo by: Csmeadphotography

WHERE WE WORK

The fireground is a high consequence, time compressed environment where we are expected to act decisively with limited information. The information we do gather must fight through ambient noise and distractions of flames, pumps, saws, and breaking glass. Add to this, masks that make it difficult to hear clearly, radios that don't work as well as we want them to, and the reality that we're often task saturated. All of these external influences increase internal stressors, creating a potent recipe for auditory exclusion, shrinking cognitive bandwidth, and temporal distortion. To say there is potential for miscommunication on the fireground is a lie, there will be miscommunication and missed communications on every fire ground. All these factors highlight the importance of clear communications during an emergency, and the pitfalls of poor communication.

How we communicate can either add clarity or cause confusion among crews. When we communicate effectively, we simplify and streamline our scenes; and when we communicate poorly, we add obstacles to our operations. So how can we enhance our communication on the fireground? This question needs to be asked in every fire department, no matter how busy they are. While this discussion is much larger than just search and rescue for the fire service, we'll do our best to stay in our lane here and focus on how our terminology and communication affect our search and rescue efforts. To do this topic justice, we want to examine communication from two different angles: first, what needs to be said; and second, what doesn't need to be said.

WHAT NEEDS TO BE SAID

Let's start with a few best practices for communication. During our radio reports, face-to-face communication outside the fire building, and mask-to-mask inside the IDLH, four simple rules are helpful to make sure that we are all on the same page and talking to each other, instead of past each other.

> *"Simplicity is the ultimate sophistication."*
> ~ Leonardo DaVinci

While the above DaVinci quote wasn't referencing the fireground, there is perhaps no better time and place to demonstrate the truth of DaVinci's famous quote. There are four general rules that we advocate for during fire operations, these are called the 4C's of communication.

- Clear
- Consistent
- Concise
- Calm

CLEAR

When we refer to being clear, we mean that we should be specific and say precisely what we mean. Being vague adds confusion and slows tempo, which is the last thing we need on the fireground. This general rule applies to all speech on the fireground, whether it's the first-in report, CAN reports, mask-to-mask communications, situation updates, or maydays. While being specific applies to all communications on the fireground, the 3-Ds…

- Distance
- Direction
- Desire

…are more particular to rescue and search. really, to all mask-to-mask communications. The 3-Ds have been passed down to us, and this simple script is very useful when we can't hear or see well. Many of us have been exposed to this framework/script in Nozzle Forward, i.e., five feet, left hand turn, push", or when more practiced, "five, left, push". An example of a 3-D script during search could be "5 feet behind you on wall three, you missed a door". The order in which we say the 3-Ds (distance, followed by direction, and lastly, desire) is very particular, and is aimed at making sure that we are consistent, precise, and understood. Depending on what you are conveying and/or how experienced and cohesive your crew is, you may not need all 3-Ds, you may only need to state the distance, because the direction and desire are obvious to your crew.

Photo by: Csmeadphotography

CONSISTENT

We also advocate for consistent terminology within our organization, and our external teammates we might operate with (neighboring fire departments, EMS agencies, PD, local utility companies, etc.). Defining terms within our organization seems obvious, and it's done implicitly at almost all departments. These implicit definitions are helpful, but explicitly defining our words clears up potential confusion. When members are unintentionally talking past each other, operations and outcomes will suffer. For example, a window-initiated search might be a one room search to my partner, but to me it means that we are going to continue to search beyond the door if conditions allow. If we clearly define what a window-initiated search is within our organization, we can all be on the same page. Some other areas where we should consider clearly defining our terms are, types of buildings (bi-level, garden-style apartment, etc.), expectations of search tactics (window-initiated search, split search, etc.), removal techniques (wrist lock, knee lock, prusik drag, etc.), etc.

Lt. Aaron Fields has been speaking for years about the need for a fire service jargon. Our take on this is that we don't need to have the entirety of the fire service having common terminology and definitions (although that would be ideal), but rather, we can't have members on our firegrounds having differing definitions or terms. This starts with consciously defining our terms, and then making sure that everyone knows them. Defining our terms helps us clarify not only the what, but oftentimes the when, where, why, and how.

We need to define our terms, simplify our scripts, and set expectations for what should be communicated. Many departments include a definitions section in their SOGs, go in-depth on when and how members are expected to communicate over the radio, and even discuss this topic in their training. However, our citizens need this to be the standard across the country.

CONCISE

Concise means that we say what we need to say as efficiently as possible, so that each word has specific intent and a high return on investment. As stated earlier, cognitive bandwidth is limited during rescue and search. This means that including words or phrases with little or no value will cause our mental processing to slow. We need to use our words to concisely paint a picture for our crews, especially since we are typically perceiving in shades of smoke and shadow. Being surgically descriptive with our words is the best way to clearly communicate our reality. This paints an accurate picture in the minds of those listening, so they "see" what we're seeing, creating a collective clarity. This clarity and shared understanding can also have a calming effect on the fireground. As Fields states, "Adding form removes fear." This is where defining our terms, going to jobs, and training can really help crews speak, "see," and think similarly.

It might seem obvious to many of us that using unnecessary and useless words leads to cognitive gumming-up of the works (and radio). Maybe not so obvious, is what inefficient word/phrases/sentences in our communications can do to our focus. When we're operating on the fire floor or floor above, actively listening to the radio or other crews is hard enough, but once someone starts to ramble, there is a tendency to tune them out. Be succinct when you speak, don't tie up the radio or slow other's thinking and tempo with unnecessary chatter. We all know firefighters that talk too much on the radio. To consciously override this instinct, before you key the mic, take a deep breath in through your nose, pause, and think about what you're going to say; so that a couple seconds later when you key the mic, you're concise, yet thorough.

CALM

Our emotions when we speak can affect how a message is received. We can likely think of multiple examples of this in everyday life, but once

Photo by: Csmeadphotography

we add in all the stressors of the fireground, this effect is amplified. Calm and chaos are both contagious.

Emerging data is demonstrating that an increase in stress leads to a decrease in our ability to verbally communicate with each other, although the opposite may be true with non-verbal communication[4]. Since verbal communication is a component of cognition, if we calm our thinking, we can calm our talking; and the reverse is also true, if we calm our talking, we can calm our thinking. This creates an internal positive feedback loop, where calm creates calm, and chaos causes more chaos. To put it another way, we become overwhelmed when we are communicating poorly, and we communicate poorly when we're overwhelmed.

Speaking calmly on the fireground seems simple, but it's not easy. This simple task takes a lot of practice to be good at. Like most skill sets, this is a learned behavior and therefore requires prior practice under stress. This means that we should utilize our radios, definitions, and scripts during our training, and also, that we should be progressively increasing the stressors during our training. While speaking calmly on the fireground is a great rule of thumb, there are absolutely times when speaking with intentional emotion can be very beneficial. This is especially true when you need to communicate that something is immediately dangerous and that operations should be changed right away. It is the fireground equivalent of using your "dad/mom" voice. Intentional emotion can cut through all the ambient noise on the fireground to clearly communicate something important to those listening.

DOES IT NEED TO BE SAID?

Hopefully these four rules make sense and parallel your experience. To many, discussing what needs to be said on the fireground seems straightforward, but sometimes when we talk about what doesn't need to be said, it can at first, appear counterintuitive. For anyone operating in stressful, time-compressed environments with incomplete information, sometimes the best communication is when nothing has to be said, and we can get our point across without words. Through experience and training, the fire service has had this tacit knowledge for generations, but let's pull this thread and see what we can learn.

A couple common examples of non-verbal communications that fire departments have been using for years to streamline and simplify operations are: riding assignments, arrival assignments, and SOGs. These three all serve the same purpose, to explicitly state our roles and responsibilities before the tones drop, so that we don't have to have these conversations when seconds matter.

Experienced fire crews have also realized that non-verbal communication inside the fire building can save time, and potentially lives. This might be a simple push from an officer "telling" you to search a room, the search group handing off a victim to another interior crew, your officer putting the TIC up to your mask to get you to see something important, etc. Nothing needs to be said in these situations, and everyone understands each other. Non-verbal communication, when used properly can, can be crystal clear and leaves little room for misunderstanding or mistakes. Overall, the smoothest incidents that I have been a part of are almost always the quietest. This is due to crews understanding and anticipating everyone else's needs, order of operations, timing, etc. Anyone that has worked with a well-trained and experienced crew can likely echo this fireground fact. I hope this goes without saying (see what I did there?), but non-verbal communication only works when the specifics have been worked out beforehand, during training. A quarterback can't walk up to the line of scrimmage, notice a gap in the defense and give his favorite receiver a knowing nod, if the quarterback and receiver haven't already had this conversation and practiced this route many times already.

As briefly mentioned before, too much nonessential radio traffic causes us to view the radio as white noise and firefighters tend to ignore much of the poetry on the airwaves. This surely leads to members missing important information. This is the last thing that we want, or that civilians need. An additional downstream effect of this inefficient information overload is that it slows operations down, by forcing us to ask the sender to repeat themselves because it's incredibly hard for task level crews to actively listen to the radio during a search, or rescue (at least for the author). This goes beyond radio traffic and can happen with face-to-face or mask-to-mask communication as well. Case in

SECTION 3: DECISIONS

WHAT NOT TO SAY

There are certain phrases that we don't believe should be stated on the fireground. These phrases, whether consciously or not, have the tendency to slow tempo and cause us to reprioritize our operations. When we use these phrases, the end result becomes slowed searches and reduced rescues.

> *"Fully involved."*
> *"Everyone is out."*
> *"Vacant."*

These phrases aren't inherently bad, but in practice, they often lead us to identify fewer searchable spaces and reduce urgency, without good reason. Think about what happens to you and your crews when you hear "Fully involved." Do you instinctively think of a defensive fire or a building with no searchable space? How often is this phrase uttered to all responding units by the first-in company that has only seen part of the building? Have there been times when some have triaged a backup line, water supply, or RIT, over search because they got tunnel vision after hearing fully involved? If we delay our primary by too long, in practice it becomes a secondary search.

Another phrase that robs firefighters of initiative, "We have reports that everyone is out of the building." Reports from whom? Does it even matter if we haven't searched anything yet? We know that in extremely stressful situations, like say, when your house is on fire, people can stop thinking rationally. Are they sure that no one is still inside? Could their kid have had a friend over? Is it possible that a relative is spending the night with them? Have they been drinking? Could someone have gone back inside for something? Could the person on scene be lying to you? Would you be willing to bet someone's life that the answer to all these questions is no? We don't even trust ourselves enough to not perform a second search, why would it make sense that we put full trust in an absolute stranger? Firefighter Rescue Survey (FRS) data has shown that in 3% (102/3007) of recorded rescues, there were reports

Photo by: Csmeadphotography

point, many of us were taught to call out what we see or feel when we're searching (i.e., "I got a bed.", "Couch.", "Window.", etc.), but to what end? Shouldn't we reserve our limited bandwidth for critical info or the unexpected. This is an example of how the best of intentions can lead to poor operations.

While it's true that brevity is critical, it can also be true that for many firefighters, especially in novel or more hostile situations, a verbal connection to your partner/crew can be very reassuring. This could be as simple as just hearing your partner's voice, or as explicit and putting a hand on their shoulder and telling them, "we got this." This verbal calming can have huge impacts on crew confidence, operations, and outcomes.

that everyone is out. As the adage states, 'it's not clear until we say it's clear.'

Lastly, at least for now, let's briefly discuss why we don't like the phrase "vacant" being used to describe a building we haven't searched yet. This one is likely pretty intuitive and echoes the paragraph above but deserves discussion. The term vacant, means that the building is unoccupied, and many equate that to mean that everyone is out of the building. This can rob firefighters of initiative and instinct, and result in a recovery, where a rescue might have been possible. If we are trying to say that the building is boarded up or damaged, then be clear and say that.

- If you locate the fire, make sure everyone knows where it is (i.e., "Fire is in the kitchen, unable to isolate, 15 feet inside the front door to the right.")

- Due to all of particulates in smoke, your voice will travel farther, and be clearer, if you can get below the smoke…this also means that you can hear better under the smoke too.

- When you're talking mask-to-mask, say it with your chest. Be loud and proud.

- When you locate a victim, state "Victim. Victim. Victim." Repeating this three times makes sure that everyone knows what you have and lets everyone within earshot know that you might need assistance.

- Once you know where you are removing a victim to (i.e., "Victim. Victim. Victim. Coming out the garage.") or what you need (i.e., "Victim. Victim. Victim. Need parallel ladders and three personnel to the Charlie-Delta corner on Division 2."), let command know so they can get EMS, other crews, or equipment to the victim as soon as possible.

- Labeling walls inside a room (1, 2, 3, 4) might simplify direction/location with your crew.

- When we talk about coordination, we often only speak of attack and ventilation (water and air) and we forget search. If we make sure the attack group verbalizes over the radio when the fire is knocked down, we allow the search group to take a more aggressive posture (i.e., split search).

- Plan and train on simplifying communication when multiple personnel are moving a victim (i.e., "Ready-Ready-Go").

- If not already done, create and practice first-in officer interview questions (add hyperlink), so that when we arrive at a career fire, we can ask these clearly and succinctly. The goal here, is that if someone puts a gun to our heads at three in the morning, we can rattle these off without thinking.

- It might also be helpful to add our first-in officer interview questions to our dispatch scripts. This way these questions can be asked, and the answers relayed, to everyone before we arrive on scene, potentially saving us time and focusing our efforts where most needed.

- Writing dispatch scripts with our dispatchers can make sure that we're getting the most important information as soon as possible.

SECTION 3: DECISIONS

- To save time, and air, consider adopting systems that minimize unnecessary radio traffic (i.e., mobile client/push button status updates, rehearsed radio scripts, etc.).

- Some other information that might be useful to relay to other crews: stair location, pertinent layout information (i.e., "Access to Div. 2 apartment is on the Charlie-side"), primary complete, change in divisions, and need for ventilation/non-ventilation/attack/etc.

- When you train, make sure you're using your radios, your terminology, and your scripts. Incorporating radios on the drill ground will help make your radio communications easier on the fireground.

- This might sound really dumb, but we also must practice actively listening to our partner, the other crews, and the radio. All of us have missed radio transmissions or communications from our partners, this, like everything, is a skill that needs to be worked on.

- Pausing our breathing for a couple seconds can allow us to hear everything better.

- Before you speak, take a deep breath in, pause, and think about what you want to say. Everyone will be happy that you did.

- How you speak (calm v excited) effects how others will receive your message and influences how they will react. Speaking in a calm voice can cause others to calm down, and speaking excitedly can cause others to become more excited. Generally, the calmer we are, the better decisions we make.

- Empower all members to be able to speak on the radio, when appropriate. Some departments are very strict about only allowing officers to speak on the radio. While this is primarily the officer's role, in many circumstances it may enhance operations to allow anyone with something important to say, to be allowed to speak on the radio.

Again, we must define our terms and be consistent when using them. We know that even if it appears that the owners aren't currently living in the building, it doesn't mean that they're not, that others aren't, or that someone isn't inside. FRS data shows that in 3% (90/2816) of recorded rescues, the building appeared "vacant" or "abandoned."

KEEP YOUR EARS TO THE STREETS

Communication is of the utmost importance. As a fire service, we do ourselves a disservice by spending little time thinking, talking, or training on this vital skill. On the following pages we have some practical communication tips that we've learned over the years from some amazing brothers and sisters The reality is that to be good at anything, including communicating and active listening inside a fire building, we must train on it. We know that we work in stressful, time-compressed environments with limited information. We know that this stress leads to all sorts of potential problems, and how we communicate can either help or harm the situation before us. Well thought out and practiced communication can smooth these friction points, leading to enhanced effectiveness and efficiency of our operations.

Note from the author - I want to thank Aaron Fields, Kevin Housley, Ric Jorge, and all the editors for sharing their subject matter expertise and time to help make sure this article was as accurate as possible.

[1] Arble, E., Daugherty, A. M., & Arnetz, B. (2019). Differential effects of physiological arousal following acute stress on police officer performance in a simulated critical incident. Frontiers in Psychology, 10, 759–759. https://doi.org/10.3389/fpsyg.2019.00759

[1] Renden, P. G., Landman, A., Daalder, N. R., Cock, H. P., Savelsbergh, G. J. P., & Oudejans, R. R. D. (2017). Effects of threat, trait anxiety and state anxiety on police officers' actions during an arrest. Legal and Criminological Psychology, 22(1), 116–129. https://doi.org/10.1111/lcrp.12077

LARGE AREA SEARCHES
CHAD DAILEY

What are some of the "best practices" and techniques taught in today's fire service that can be applied to "large area searches", particularly while operating at a working fire incident? To answer this question, we must first have a shared definition of what a "large area" is. In the past the fire service would have, and still might, consider this term to describe the common grocery store, a free-standing department store or maybe an indoor lumber yard, present in nearly every community throughout the country.

Tactics are often applied based off our past experiences; however, we should not apply these "bread-and-butter" single family residential fire experiences and tactics to incidents in which the buildings hold almost no similarities. The evolution of buildings we are confronted with in this country necessitates the need to be more defined in our terminology. Notably, as this will often set the stage driving our tactics and decision making throughout the incident. An average grocery store in our country has around 50,000 square feet, while an average big box store (Walmart) has up to, and sometimes over, 200,000 square feet. Now, we are facing more and more buildings across our country that can have up to 1,000,000 square feet under one roof! That is a "large area search", or more like "searching for a needle in a haystack". These are the types of buildings where we must consider our approach and formulate a tactical plan, before the bell rings and the call comes in. Much like the time and energy many departments will invest into "high rise" firefighting skills, tactics and equipment. Even though the rate of occurrence may be low for these large structures, our risk factor is significantly higher should we have an incident of substance in one of them.

WHERE THIS COMES FROM

My perspective on this topic is derived and influenced by much training with both inside and outside resources and through diverse experiences and circumstances throughout my 32-year fire service career. In addition to this, some unique characteristics of Kansas City, Missouri, not commonly found anywhere else in the country. Incidents of this magnitude weigh heavily on my perspective and opinions.

By some accounts, Kansas City, Missouri has more mined-out caverns (subterranean commercially used space), than anyone else worldwide. These contain roads and infrastructure that service hundreds of individual businesses. In addition to this, we have been home to multiple automotive plants, one of which contained 4,900,000 square feet, and had been a building protected since 1929.

Now, like other departments, we are faced with mega warehouse and distribution centers including recent million square feet meta data centers.

Battalion Chief John Tvedten, a 26-year veteran of the Kansas City Fire Department, passed away December 18, 1999, in a large cardboard manufacturing warehouse after conditions rapidly changed. He ultimately ran out of air before being able to gain his orientation and exit the structure (NIOSH Report #F99-48). This experience preempted our department to look at and consider our approach to such incidents. From Incident Command to search tactics, we looked at ways to mitigate incidents more effectively in structures of this type and magnitude.

Challenges we've been faced with have not been small. For example: Incident Command, size up and building construction, communications, fire fighter skillsets and training, equipment needs and codes enforcement. All these things play a huge part in the outcome of such infrequent incidents as these. "Large Area" incidents are not an everyday occurrence for most of the American fire service. Historically, many large industrial spaces were protected by specially trained employes (fire brigades) that would mitigate or limit the need for a local fire department response. Often these will be federal, state or local government installations where we are restricted or even prohibited to enter (i.e., a very large, multi-story U.S. postal hub or a complex that used to be a factory for manufacturing components for the US nuclear weapons arsenal). These are buildings which we were barely able to enter and influence code enforcement much less, preplan for a large area incident.

IT STARTS WITH COMMAND

It's imperative that a well-rehearsed, well-trained incident command system be used in an event of this nature. Without this in place, the potential for failure is almost certain. All members involved must know how to work, function, and communicate within the department's chosen incident command structure. Any freelancing could significantly break down and hinder operations and cannot be accepted by even the most senior members. A chosen model of IC must be taught, starting with cadets. It must then be reinforced continually with field personnel. Considering these massively large buildings, it is necessary that the chosen IC model is scalable, and commanders well versed in utilizing it to its full potential. To teach and implement such a program, and go years without any continuing education, will be detrimental. This type of incident is not a typical " house fire" with pre-determined tasks where firefighters can show up and jump into action. Many factors can and will affect the complexity of the IC system with an incident involving these types of structures. Current needs should be assessed with the understanding that additional resources will be called upon to appropriately staff and manage this type of incident.

There are various roles needed to support a command structure for an event of this magnitude (i.e. suppression, ventilation, water supply, search, communication, EMS, Haz-Mat, logistics). The duration of the event, weather, time of day, purpose of the building and its contents can all add stress and additional needs that will have to be addressed through command structure. Delegation of responsibility, as established in most IC models, will be paramount to success in mitigating such calls and the making of sound decisions. Keep in mind that these incidents may involve multiple agencies, which is another reason for a strong IC system that is known and practiced by all agencies involved. This would, and should, encourage multi-agency training.

SIZE-UP CONSIDERATIONS

Actions and mindset from the time of receiving the initial alert to arrival could make or break the operation with these mega structures. In some ways, it could be treated like a "high rise" incident, in many other ways, there will be no comparison. This is an incident where the most experienced IC staff should be working and not a "training day" for newly promoted people. Additionally, these buildings are showing up in smaller communities where departments haven't had to implement "high rise" planning and techniques. This means size up must be inclusive of so many variables: building construction, overall size, layout, infrastructure controls, points of access and egress, contents, operations within, and so much more. Simply knowing the construction type and layout of the building can help determine combustibility of the structure, how it will affect fire and smoke spread, and the hazards firefighters will face in searching for and mitigating a fire within it. It will also determine if walls can or cannot be breached for closest access points to the incident, if no natural entry point exists. The roof should be a very good source to include in overall size up, along with often being the best vent point for this type of fire. It could potentially be a quicker way to locate the area of suspicion. Drones and TICS are technology that needs to be considered early on for gaining a well-informed size up,

provided availability and competency on these tools exist within the department. The overall size of the structure and how, or if, it's divided up needs to be learned early on, if not already known in a pre-plan.

Suppression systems, ventilation systems, fire walls, curtain doors on fusible links, working machinery, self-guided machinery, hazardous materials, and other things specific to these structures need to be known and considered. These systems can significantly enhance operations, and if not used properly can sometimes hinder operations or even pose life safety threats to personnel. Consideration for multiple teams working from different access points should be made in searching for life and fire in a structure of this size. The "brain center" of these structures may give vital information like floor plan, fire doors, live camera feeds, location of alarm activation and specific sprinkler zone activation. Never disregard the employees and/or building maintenance personnel that know these buildings intimately. Keep them someplace easily accessible or assign them a" custodian" if necessary. It's my experience that additional questions for them will arise for them as the incident progresses.

COMMUNICATIONS & ACCOUNTABILITY

The ability to communicate with various personnel and teams will be critical with an incident that stretches out over an area of this size. It's likely that more than one channel will be needed as the incident accelerates and is scaled up. Communication will also be critical in the ability to maintain some kind of accountability of personnel while working on different tasks. Likely, assignments will need to be made in multiple areas throughout the structure. Anything that can enhance or hinder whatever radio system that is in place, i.e., repeaters, Wi-Fi features, cell phone applications, and even hardline communications are things that need to be considered, and hopefully preplanned for. Know the capabilities of the radio system in these structures before this day comes.

Accountability is often very reliant on the equipment and software chosen to communicate with and, though there is new and actively emerging technology on the market to help with this, very few have proven to be reliable, especially if altitude (multiple floors) is added to the equation. Flying drones indoors with light smoke, short wave IR markers and cameras, breadcrumbs of hot and cold packs for the TICs, unmanned robots, large trailer mounted fans and battery powered vehicles are just a few of the things currently being used and experimented with to apply to a fire in a structure of this size.

FIREFIGHTER SKILLSETS

This is where the biggest investment of time and energy must be spent to have a successful outcome with these incidents. A well rounded, forward-thinking, confident and motivated group of firefighters who have invested time into the skills that come into play in such an incident will yield large dividends. These skills typically will not have changed much, but the application could be far different from previous training and experience.

Whether it is a rope assisted search, use of thermography, forcing doors in zero visibility, breaching walls, making extra-long stretches, knowing how to pump a building in a known district, very long drags or carries with boards or other techniques, a multitude of skills will be required to mitigate this type of incident. With the influence of other departments we contacted, and a lot of hard work from our own personnel, we came up with our own version of a rope assisted search, the Kansas City Large Area Search Team (L.A.S.T.). This was implemented following our LODD in 1999. This is a skill that must be trained regularly and will get personnel 300-600 feet into a building. This technique requires keeping it simple, and continual training is essential to its success. IT has been adapted to large area incidents in the past but start

SECTION 3: DECISIONS

and access points must be strategically picked. Entering a zero-visibility environment on a rope requires skill and a constant awareness of the surrounding environment. These large spaces often have moving machinery, uneven floors with holes, overhead obstacles, and many other hazards specific to the building that firefighters won't be able to see. Whether it's an Amazon packaging building, an automotive assembly plant, a bottling plant, etc., each one will be unique with its own hazards.

EQUIPMENT NEEDS

This is where the department needs to determine the best way to increase capabilities of what is already in place. Thoroughly vet any ideas of new equipment or upgrades. Perform hands-on practical use of any new ideas for equipment and techniques, or don't commit to them. Equipment and ideas need to be vetted in the environment and atmosphere in which they will be used, preferably on site for some of them. This applies to all things, such as: coms, TICS, tracking equipment, rope assisted search, larger SCBA bottles or rebreathers, and so many other things being marketed to the fire service. Transfilling or carrying extra SCBA bottles may prove beneficial depending on the task at hand and the conditions encountered. RIT functions will be challenging, and multiple teams will be needed for a structure of this magnitude. Ideally, this would be planned for from the onset of the incident if resources will allow for it. Equipment to transport firefighters, sometimes in smoky environments, throughout the structure often gets overlooked until such an incident occurs. The use of electric golf carts has proven to save much time and energy, in our experience. Additionally, very large ventilation fans and fan boats have provided ventilation at such incidents, with good success.

CODES & PREPLANS

There is really no excuse for not having performed walk-throughs and pre plans for buildings of this size and scope within a jurisdiction, unless legally prohibited. A pre plan can prepare everyone involved for overcoming the enormous amount of ground and space that may have to be covered to carry out an operation in a structure of this size. A pre plan should clue one into all things discussed earlier. Notes should be meticulously taken and added to the CAD for reference upon receiving a call at the location. Codes can be of great help to not only prevent or rectify hazards the occupant may create. During the construction of such buildings, we can request things that will enhance our abilities: fusible link sky lights, radio repeaters, appropriate pressure of the fire pumps, IR markings on pillars or walls, and many other things that will sometimes be accommodated. It is also possible to have special requirements for buildings of this size to be written into a city's own ordinances. National fire code already requires sprinkler protection, defibrillators, emergency shut offs, emergency exits and lighting and fire extinguishers, but there are many more to consider. These requirements have been altered through the

years as buildings and hazards have changed and been recognized. We must educate all of those involved and bring attention to the inherent dangers and challenges we will be faced with when a fire breaks out in a building of this size. It's not far-fetched to think that some of the special needs for an incident at a large structure be imposed on the owner. Much like we have worked with owners to provide repeaters/routers, electric vehicle transport devices and more. Other departments have made requirements for special ventilation fans on site, rebreathers, hose cabinets with 2.5 hose and street pipes. We need to step up awareness on every level of the organization and get out there and require the community to allow us to be proactive in the job they expect us to do. These needs and concerns should be addressed through the fire marshal's office at the time of inspection. These mega structures likely need to be looked at within the National Fire Code that many cities adopt. There first needs to be evolving additions and amendments for these such structures, as we are still in the learning process of how to mitigate such incidents. Our lives and the lives of those we serve depend on these codes, pre-plans, and the collaboration of all levels of the organization, with the public and officials governing it.

IN CONCLUSION

We must take a holistic approach to an incident in a structure of this size and magnitude. To perform a "large area search" in a structure of this size is a monumental task, given the conditions we could likely be functioning in. This task of clearing the building along with mitigating whatever is happening inside the building will likely stretch resources and abilities of personnel, beyond that of a typical incident. A robust command system from the start is going to greatly help ensure a successful outcome of such an incident. The approach to successful mitigation of these incidents needs to come from more fronts than just one. We must open the minds of lawmakers to enable fire marshals to impose requirements needed on these structures as they are being built.

We must be open-minded to new techniques and equipment that would enable us to work safer in these large areas. We need to train at all levels within the organization for the tasks that will be performed at such incidents. Robots, drones, IR devices, rebreathers, electric mobility devices, different types of communication devices - all things need to be considered with an open mind. We need dedicated firefighters not afraid to put ideas to the test and invest sweat equity to prove new techniques and equipment for this purpose. These experiences then need to be shared with the rest of the fire service. There have been, and will be, failures in our efforts and we have had many. Learning from failures is indeed progress, provided pride doesn't get in the way.

As I wrote about previously in this article, we recently had a search and rescue operation of a mined-out cave with two people lost inside. Middle of the night, zero visibility, and uncommercialized space. The cold pack "bread crumb" idea I shared with others on the job was implemented to this operation with success. The people assigned to the incident had trained extensively on rope searches along with other large area search techniques and were very well versed in thermal imaging. Ultimately, they found the two lost parties and had them out within a couple of hours by utilizing multiple skills, several hundred feet of rope, and then ice packs with TICs. The IC called for the use of hot/cold packs to extend their search beyond that of the rope. Go, train, influence, and take risks. Keep an open mind to new techniques, new technology, and unconventional ways to accomplish an uncommon task. Be willing to be uncommon in your mindset and approach, knowing this makes us vulnerable.

THE IMPACT OF ERA ON SEARCH

JAMES JOHNSON

Photo by: *Csmeadphotography*

When discussing residential occupancies in the context of search and rescue, there are many individual variables that come into play. Geographical area of the country, level of affluence, era of development, and population density all play a distinct role in the types of structures we may encounter. While there are many similarities and commonalities within residential occupancies from coast to coast, many large cities have unique housing challenges that they have adapted to. For instance, in Vancouver, British Columbia, and in many other cities across North America where housing prices have skyrocketed, it is not uncommon to find traditional single-family homes, divided into four to eight individual occupancies. These alterations create challenges when it comes to navigating interior spaces, due to the level of compartmentalization and unnatural layouts. As a result of this change, keeping a watchful eye for additional electrical and gas meters, individual suite numbers, and multiple doorbells and mailboxes, should become an important part of our initial size-up. This is just one example highlighting the importance of localized knowledge of the types of occupancies that are found in your districts and cities and stresses the fact that there is no substitute for getting out and walking your areas

E	=	ERA
M	=	MATERIALS
U	=	USE

EMU

A number of years ago, I started to develop a personal system of building size-up that I began to use at fires. Over time this has evolved, and with the input from colleagues, it has been refined, resulting in the acronym EMU (Era, Materials, Use). Initially, this was used to solely identify the building construction, but as I started using it regularly, it began to help paint a wider picture of the structure as a whole and really aides in understanding many facets of the building. For the purpose of this article, and how it relates to search, I am going to focus on Era and Use.

While this size-up model is very useful, it should always be done in conjunction with a comprehensive size-up of the physical characteristics of the structure. Door location and swing direction, window size and shape, chimney location, vent stacks, and other exterior markers can very accurately help identify and prioritize searchable spaces within the structure and help to create a mental road map of the interior floor plan.

Jordan Legan, a firefighter in Seattle, Washington with his master's degree in architecture from the University of Washington, leads one of the most thought-provoking classes that I have taken. In his class, 'Wayfinding in Smoke', he discusses a couple of concepts that really stuck with me and changed the way I view a building when attempting to orient interior spaces. The first concept that he talks about is that in most architecture there is a clear delineation between the public and private sections of a residential occupancy. This could either be a horizontal separation in a one-story house, or a vertical separation in a multi-story private dwelling, yet there is usually a clear and thoughtful divide between these two spaces. The second concept he discusses is the common arrangement that stacks the kitchen, dining room, and living room in the public side of the home. These two concepts can be incredibly beneficial in aiding firefighters to identify and prioritize searchable space within a structure.

ERA

I firmly believe that from both a building construction and size-up perspective, era is the most important characteristic that we need to identify. Era has a direct correlation to construction methods, materials and structural systems, which has a downstream effect on how the building is laid out and compartmentalized. Older structures tend to be far more compartmentalized due to the available materials at the time, whereas the development of lightweight materials and systems provided the ability to create larger compartments, and lead to the introduction of open-concept design.

When we look at the era of the building, we can break down buildings into three categories. Pre-war, Post-war, and Modern. After an extensive review across North America, these categories have been found to cover the major milestones in building construction and technological advances spanning the continent. While the buildings that fall into these three categories may look slightly different depending on what part of the country you are in, the significant changes in the way they are constructed and how that impacts the interior layout remains fairly consistent.

PRE-WAR: UP TO 1940

The pre-war era of building covers the largest time span and contains some of the most diverse displays of architectural stylings. Although the individual architectural style can aide us in identifying the era, the differences in interior layout and compartmentalization is far less drastic than the differences in exterior appearance. Due to the early age of these structures, and the technology and resources available at the time, the interior layout is very compartmentalized.

Pre-war homes can range anywhere from early small single-story structures, all the way to large ornate two and a half-story homes.

An important part of the initial size-up in this era of homes is identifying occupied space within the half-story. This can be accomplished by identifying windows within the gable ends of the roof, or the presence of dormers in the roofline. Due to this era of homes being constructed using balloon-framing, it allows a clear path of travel vertically within the wall assemblies and fire extension to the top levels can happen very rapidly.

COMMON STYLES ASSOCIATED WITH THE PRE-WAR ERA

- Victorian
- Queen Anne
- Bungalow
- Craftsman
- Tudor
- Cape Cod.

Common building characteristics: balloon-framing, actual dimension lumber (wall, floor, and roof assemblies), shiplap sheathing and subfloor, lath and plaster interior coverings.

PRE-WAR ERA SINGLE FAMILY RESIDENCE

COMMON STYLES ASSOCIATED WITH THE POST-WAR ERA

- Split-Level
- Ranch
- Minimal
- Traditional

Common building characteristics: Platform-framing, nominal dimension lumber (wall, floor, and roof assemblies), introduction of plywood sheathing and subfloor, introduction of wood paneling and gypsum interior coverings.

POST-WAR ERA SINGLE-FAMILY RESIDENTIAL

SECTION 3: DECISIONS

COMMON STYLES ASSOCIATED WITH THE MODERN ERA

- **Contemporary**
- **Ranch**
- **Craftsman**

(Note: The modern era of residential occupancies draws a lot of design inspiration from homes of past eras. This highlights the need to utilize other aspects of our size-up to help identify the age of the building. The timeframe that the area was developed, modern exterior finishes and trims, as well as modern doors and windows, can assist us in distinguishing between an older era of home versus a modern rendition of a classic style.)

Common building characteristics: Platform-framing, nominal dimension lumber, engineered lumber, lightweight floor and roof assemblies, OSB sheathing and sub-floor, and lightweight gypsum interior coverings.

MODERN ERA MULTI-FAMILY RESIDENTIAL

POST-WAR: 1941 TO 1975

The post-war era of North America saw one of the largest influxes of buildings in its history. Soldiers coming home from WWII and the Korean War needed a place to call home, and government legislation, such as the Federal Housing Administration and the Serviceman's Readjustment Act, helped to drive the building boom. Due to the

Photo by: *Csmeadphotography*

necessity, large demand, and shortage of construction materials, the homes of this era tend to have a very basic and functional design. Often boxy and ordinary, these simple homes are typically one or two story and relatively small in footprint, with many being under 1000 square feet. Just like the single-family homes of the time, the multi-family dwellings such as townhouses/row homes, and apartments found in more urban centers, also have a utilitarian or plain appearance in comparison to their pre-war counterparts.

Similar to the pre-war homes that came before them, the available materials and building technology of the time resulted in interior layouts that were very compartmentalized with small individual living spaces. One common feature to this era is the incorporation of garages and carports attached to the home, as it was during this time period in history when it became commonplace for many households to have an automobile.

MODERN: 1976 TO CURRENT

The modern era covers the period of time from 1976 up until current day. I think a lot of firefighters might be surprised to see the mid-1970s as the starting point for the modern era, but there were a lot of technological advances in the construction industry that began to make their way into buildings during this time frame. Floor assemblies started to incorporate wood I-joists and parallel chord trusses, allowing the clear spans to increase, and resulting in larger interior compartments. The use of lightweight roof trusses further expanded the ability to create large open spaces, which assisted in the development of the open-concept design.

Although there have always been examples of large homes, or 'mansions', in affluent areas within all of the era categories, the modern era really brought forth the concept of the 'McMansion' home, and lightweight assemblies made them much easier to build. These homes are often 4000 to 10,000 square feet and can have features that are not commonly associated with many other house styles. These features can include large walk-in closets and spacious ensuite bathrooms that can become disorienting during search.

USE

During the initial size-up using the EMU model, the 'U' stands for 'Use'. In the context of fire ground search, this is where we are trying to identify the potential life hazard of the building. The first thing we need to determine is if it is residential, commercial, or residential over commercial.

If residential, then we need to identify whether it is single-family or multi-family. While most multi-family occupancies come in the form of either townhouses/row homes or apartment buildings, as I mentioned earlier, it is not uncommon in urban areas to find what would traditionally have been a single-family home converted into multiple individual dwelling units. These converted homes tend to be quite disorientating as they typically have unorthodox layouts and compartmentalization. It is not uncommon to find bedrooms without exterior window access if the remodels were completed illegally and without permit.

In addition to the use, our search operations may also be impacted by the era of the building, the types of structural materials, and the involvement of the fire. In the high-rise setting, due to the non-combustible nature of the structural materials, it may be prudent to have most occupants shelter-in-place and prioritize those occupants located directly above the fire compartment or on the topmost floors. Whereas in the low-rise setting, particularly in modern era multi-family buildings, the potential for horizontal fire spread is much greater, and we may need to shift our focus to include adjacent units in addition to those located above the fire.

INGRAINING MENTAL MODELS

When we break down building size-up into these simple categories, it becomes quite easy to train a firefighter to look at a building and be able to identify the difference between these era ranges, and it really sets the foundation for being able to anticipate the construction materials, as well as provides an understanding of how the interior spaces may be oriented. I have found that real estate websites can be a very useful tool to help establish common themes and visual cues for the different eras. Most real estate listings include the year the home was built, and when we pair that with the exterior photographs, it can assist us in creating our mental models for each era range.

GAINING INSIGHTS

Dr. Gary Klein is a cognitive psychologist and has spent a large portion of his career researching how people make decisions. He is most notably known for his work in the area of recognition-primed decision making (RPD). One area that Dr. Klein has studied is how the process of gaining insights is the best way to improve decision making. He breaks down the process of gaining insights into three simple pathways: a connection pathway, a contradiction pathway, and a correction pathway.

CONNECTIONS

A connection pathway is when we see something that aligns with our beliefs or our existing mental model on a given subject. To use an example from building size-up, this would be a firefighter looking at a building from the exterior, making an assessment on the era that it was built, and then during the process of the fire fight, that initial

Photo by: *Csmeadphotography*

assessment is deemed to be correct. Additionally, they may discover other connecting features that further reinforce their existing mental model.

CONTRADICTIONS

The next pathway is the contradiction pathway. This is when we encounter something that is at odds with our initial assessment. An example of this would be when a firefighter sizes-up a building and determines that it is a pre-war home. Upon making entry, they encounter a wide-open floor plan common in modern open-concept design. In this example, the firefighter is experiencing a modern building that was built to match the design aesthetic of the neighborhood and highlights the fact that we need to be constantly sizing-up and taking in new information during the entire operation. This contradiction to what is expected causes the firefighter to re-evaluate their initial assessment and adjust. While this gain in insight doesn't necessarily change the firefighter's initial building size-up mental model, it causes them to investigate further and brings the awareness of new possibilities.

CORRECTIONS

The last pathway is the correction pathway. This is when we become stuck because of a flawed belief, and the only way to move past this flawed belief is to become curious, investigate the subject further, and ultimately change our mental model. I believe that there are many of us, myself included, that have experienced this correction pathway as a result of the work that has been done by UL Fire Safety Research Institute to provide understanding of the science behind what we are experiencing in fires.

This is the method of gaining insights that many organizations struggle with the most. Dr. Klein explains how many organizations believe they want insights and want innovation, but in reality, they don't. Gaining insights is disorganizing, as they make you change the way you think, and can upset the balance or the processes within an organization. As firefighters, and as an industry, we are notorious for holding onto our beliefs, and I'm sure that everyone reading this has experienced an apprehension to change our preconceived ideas on a given subject. Although it can be disorganizing, the correction pathway can be a powerful way to gain insights and improve performance or decision making.

INSTINCTS & TRAINING

I firmly believe that the work of Dr. Klein directly correlates with the process of gaining insights during the size-up of a building. If we keep these insight pathways at the forefront of our mind, we can be more likely to reinforce our existing mental models, be able to identify when we encounter contradicting information, and lastly, be willing to break free of flawed beliefs that are holding us back from further insights. Size-up and identification of the building should be ongoing throughout the incident, and it's important to trust our instincts when we start to encounter characteristics within a building that are different from our initial assessment.

HIGH-RISE SEARCH & RESCUE

BRENT BROOKS, JIMMY DAVIS & CLARK LAMPING

Search and rescue operations in high-rise buildings have always been challenging, even for the largest departments. A fire in a high-rise building yields two concerns that must be addressed quickly: the fire itself and the welfare of building occupants in harm's way. The three most common occupancy types of high-rise buildings are residential, commercial, and hotels. Each of these occupancy types will require a different search and evacuation strategy. When discussing search and evacuation efforts in high rise buildings, we need to understand that occupancy will dictate strategy, strategy will dictate our tactics.

TRADITIONAL SEARCH

The traditional search method commonly used for residential houses known as a right or left-hand search is ineffective in a high-rise building as it may take hours to search a building with several hundred units. Considering the number of occupants during a fire is important because they significantly outnumber the available firefighters. Rescuing just one person can require 2 to 4 firefighters, making it challenging to balance the need for rescue, containment, extinguishment, and evacuation. Focusing too much on one aspect can negatively impact the others.

"Search and rescue efforts required for a 100' x 200' foot floor area of a high-rise office building could have 150 small rooms, cubicles, and spaces that must be searched can take 2.5 hours to search for a team of 2 firefighters to complete the search"[13]

In high-rise buildings, our greatest concern is vertical smoke migration through floor penetrations, utility chases, elevator shafts, interconnected HVAC ductwork, gaps between the building frame and the exterior curtain wall, and stairwells. When smoke spreads in a building, it becomes a bigger problem as it invites more people into the emergency. This, in turn, increases the search areas for firefighters, which requires more resources to address the situation.

TWO-WAY LIFELINE

The stairwells are our lifeline in and out of the building and, in most cases, are connected to all other floors and compartments. Both firefighters and occupants need to use them. Once the stairwells are filled with smoke, occupants in the stairwells become incapacitated. Firefighters in full PPE, working in zero visibility, become fatigued and move slower. Most fatalities at high-rise fires are result of smoke asphyxiation from victims who were trying to evacuate via hallway corridors and stairwells. These victims will be found throughout the building, even in areas remote from the fire. In the Cook County building fire, (2003) 6 occupants were discovered deceased in stairwell because of locked stairwells. At the MGM fire in Las Vegas on November 20, 1980, the majority of the 85 fatalities we located in hallways and stairwells inside the tower even though the fire never extended beyond the first floor. Stairwells remain a high-priority search at any given high-rise fire. Building systems designed to pressurize the stairwells provide us with more time to evacuate occupants and limit the smoke movements into compartments during fire attack. Positive pressure and isolated stairwells can work as our invisible allies in such situations. Firefighters can use fans to create positive pressure in stairwells that are not equipped with the systems. However, we face some limitations depending on the building's layout and size.

EXPECT CHAOS

During a fire, the fire alarm should activate, and the public address system may advise the occupants to evacuate the building, sending hundreds, possibly thousands of panicked residents out in common areas such as hallways, stairwells, and elevator lobbies. This massive evacuation effort

will congest the stairwells, create confusion, panic, and delay the firefighters who are using the same stairs from getting to the upper floors with our equipment. The fight-or-flight instincts that humans experience in stressful situations can lead to unpredictable behavior. Those who need to evacuate during an emergency must be provided with a clear and safe evacuation route. Meanwhile, those who are sheltering in place require protection or rescue. However, this can be challenging since there is often more ground to cover than firefighters can manage on foot. Climbing stairs, wearing heavy bunker gear, carrying equipment, and relying on SCBAs can all make it difficult for firefighters to move quickly and efficiently. If the building's system is capable, we can keep occupants safe during a fire emergency by using the public address system to instruct them to stay in place or evacuate through a clear route. Instructions to building occupants should be short, concise and action oriented. Be mindful of language barriers, it may be necessary to utilize a translator when making building announcements.

SHELTER-IN-PLACE CONSIDERATIONS

A shelter-in-place strategy is ideal for high-rise fires. Room doors that open into the hallway are, by code, solid-core fire doors equipped with self-closing hardware. Occupants who remain in their rooms will be safe for an extended period of time. A shelter-in-place strategy will also keep the stairwells clear, allowing the fire department to gain access to the upper floors without dealing with evacuating residents. A full shelter-in-place strategy would also allow firefighters to strategically and deliberately search for and remove occupants in an organized fashion, starting with those who are in the most danger. An exception to the shelter-in-place strategy would be an exterior fire extending vertically up the sides of the building, as we witnessed at the Grenfell Tower fire in London, England in 2017.

TRADITIONAL PRIORITIES

The priorities of searching high-rise buildings are the same for residential buildings- to immediately identify, locate, and remove the civilians who are in the most danger. Evacuation of occupants on the fire floor should not be attempted until the fire has been extinguished and the smoke has been completely removed from the floor. This could take an extended period of time. Before the fire door is opened, all occupants must be removed from the floor or directed back into their rooms to shelter in place until they are notified by the fire department that it is safe to evacuate. The units on the fire floor should remain smoke-free if their doors are not opened. Once the smoke lifts, a secondary search is completed, and wellness checks are done on all other units with a carbon monoxide detector. Once the rooms directly above the fire room have been cleared, the search and evacuation should extend to the entire fire floor, then the 2 floors above. These 3 floors are the most critical; these people are in the most danger.

PRIORITIZE AND EXECUTE

Now that we understand building science better and have studied our past fires, we no longer search closed doors during the primary search. It takes considerable time to force open doors and search every unit during a rescue operation. This can slow down or even prevent a thorough search from being conducted. Firefighters have limited time to complete their search before they are required to exit due to their depleted SCBA bottles.

Conducting a common area primary search generally involves three firefighters using a thermal imager to search the hallways rapidly. They move from one end of the hallway to the other, shoulder to shoulder, as fast as they can walk. This technique is called "snow plowing". Firefighters can bypass closed doors to prevent the occupants from exposure to toxic smoke. Crew integrity must be maintained at all times.

> **Search and evacuation priorities:**
>
> 1. The fire unit
> 2. The common areas on the fire floor including hallways and lobbies
> 3. The stairwells
> 4. The units directly above the fire unit
> 5. The common areas of the 3 floors above the fire floor including hallways and lobbies
> 6. The common areas on the uppermost floors, beginning at the top floor including hallways and lobbies

CONSEQUENCES

During fire suppression and evacuation efforts, smoke will inadvertently enter the stairwells and begin to accumulate at the top. Any occupants using the stairs for evacuation could find themselves in an IDLH environment without respiratory protection. A high-priority search must be extended to the designated fire attack stairwell prior to fire attack. It is suggested that fire attack be withheld until the stairwell is searched a minimum of five floors above the fire floor. Occupants encountered in stairwells must be relocated to safety before commencing fire attack and subsequent smoke contamination of stairwell. Stairwell searches must be made prior to, during and after fire suppression efforts. Stairwells must be given an absolute high priority in terms of a good search. The next priority for search and evacuation is the uppermost floors, starting at the top. It is here that smoke will be accumulating. Deploying valuable resources to the very top floor may not be feasible. Smoke travel is dependent upon variables such as: location (floor) of fire, height of the building, inside and outside temperature differentials, and location of fire relative to stairwell and elevator shafts. Chicago Fire Department (CFD) employs a Rapid Ascent Teams (RAT) – which are dedicated task-level assignments designed solely for the intent of stairwell searches, assessing smoke conditions on the floors above, and occupant well-being checks. RAT will provide valuable CAN reports. If the fire alarm has been activated, and the public address system has advised occupants to evacuate the building, people will be found in stairwells, hallways, and elevator lobbies.

These people will be people attempting to use the elevator for evacuation, waiting in an unprotected lobby, pressing the button for an elevator car that is not coming because it has been locked out in the lower lobby by the alarm system.

IN CONCLUSION

High-rise firefighting operations are very complex, but we can find solutions by studying past fires and researching different perspectives. Ensuring fire safety requires a collective effort from firefighters, building occupants, operators, engineers, designers, city planners, and architects.

---

[13] (Dunn, 2022 Skyscraper Battlespace High-rise Firefighting.)

OPTIMIZING VENTILATION FOR SEARCH

NICHOLAS PAPA

Photo by: *Csmeadphotography*

Possessing a practical understanding of fire dynamics allows ventilation to be leveraged to its full potential. By enhancing tenability within the affected areas, thereby facilitating our operations, we can maximize victim survivability, as well as property conservation – fulfilling our core mission. While the success of ventilation has always required a coordinated effort, the evolution of the fireground has made it more imperative than ever.

IT'S ALL ABOUT TIME

Because of the increased fuel loads being encountered, mainly comprised of synthetic materials (i.e., "plastics"), todays fires have become significantly fuel-rich. The imbalanced air-to-fuel ratio creates ventilation-limited fire conditions. The inefficiency of the combustion process causes the fire to burn "dirty," leading to an increase in smoke production and toxicity – reducing visibility and tenability. The synthetic fuels, additionally, have double the heat capacity, and four times the release rate than those naturally derived (e.g., wood, cotton, wool, etc.); accelerating the timeline of fire growth and spread by a considerable measure. Today's fires have all the necessary heat and fuel, but they are starving for the oxygen leg of the triangle. While promptly removing the smoke is a top priority, the responsiveness of the fire to any increase in fresh air, prior to suppression, has severely reduced our window of opportunity and our margin for error; we need to be more precise and proficient than ever before.

TEMPERATURE AND PRESSURE

Ventilation is principally a matter of temperature and pressure. Heat rises and gases expand. The expansion of the gases is what generates pressure. Fire and its byproducts seek to

Photo by: *Csmeadphotography*

move from high pressure to low pressure, flowing away from the seat of the fire and taking the path of least resistance along that gradient, to upper floors, adjacent compartments, and the exterior, via stairways, doorways, windows, and other natural openings. This movement results in an inverse reaction with the fresh air inside the structure and from the outside, causing it to be drawn towards the seat of the fire.

The mutual dependency of both the intake and the exhaust components are most fundamental to the function of ventilation. In order to achieve and sustain the desired flow of fresh air in, exchanging with the byproducts of combustion out, the inlet and outlet openings must be complementary. A basic analogy for this relationship is "shotgunning" a beer. When the can is typically opened at top, using the tab to penetrate the perforated section, and is tipped over, the beer slowly sputters out. Because the hole is serving as both the inlet and the outlet, the two functions competing against each other for the same space, the bidirectional flow is choppy and inefficient. By tipping the can over and piercing the side of the can before popping the top, the additional opening becomes a dedicated outlet, allowing the hole in the top of the can to function solely as an inlet— creating an optimal unidirectional flow. If the openings are not of comparable size, the smaller of the two becomes the limiting factor; ensuring they match allows for optimal flow.

INTAKE IMPACT

The exchange of fresh air with the byproducts of combustion— allowing the smoke to lift and the environment to lean out (the fire burning more efficiently and producing less smoke and more flame)— is only taking place between the fire and the available openings, which means *the benefits of ventilation are only occurring within that space*. No matter how much the involved area is vented, horizontally or vertically, the relief will only be experienced along the intake pathways— with the greatest impact being closest to the inlet opening, predominately the main entry doorway, and along the floor.

The rooms remote from those pathways, even if interconnected, however, will remain largely unaffected. Unless an open window or an exterior door is present within those distal spaces, there will not be an inlet to pull a draft through them. Without the intake of fresh air, there will be no exhaust of the byproducts of combustion, and therefore, no improvement in conditions – *no exchange, no relief*. A simple example of this concept is demonstrated in the following childhood trick. When a straw is placed into a full drinking glass, the liquid will flow up through the straw. If the straw is pulled out, the liquid remains in the glass— fresh air intaking through the top of the straw and replacing the liquid as it exhausts out of the bottom of the straw. By placing your thumb over the top opening— occluding the inlet— the liquid will remain inside as you remove the straw from the glass. It is not until your finger is lifted off, uncovering the opening and allowing air into the top of the straw, the liquid is able to flow out of the bottom.

Photo by: *Csmeadphotography*

ELEMENTS OF VENTILATION

The elements of fireground ventilation are the location, the size, and most of all, the timing of the openings – the true lynchpin of the operation. Successful coordination requires adherence to the following principles: control, of both yourself (maintaining composure and positional discipline) and the environment (managing the openings to "make it behave"); communication (verifying the progress of the interior crews); and competence (in the mental, physical, and tactical domains). Because ventilation affects the environment, and therefore, all other operations, as well as any trapped occupants, it has strategic implications. Before creating any openings, the purpose and the consequences of doing so must be identified, as doing so can single-handedly alter the course of an incident. There are three objectives for performing ventilation: venting for extinguishment; venting for search; and venting for access.

Often not recognized as a form of ventilation, first action taken on the fireground is typically opening a door to gain entry. This is also the case when breaching walls and ceilings to access void spaces for concealed fire. In both situations, a nozzle team must have a charged handline in position and be ready to advance/operate when the openings are created; preventing the intake of fresh air from causing undue fire growth and spread. If a crew is ready to initiate the primary search before that time, however, they should not wait, if conditions allow. To proceed, the search crew must simply control the door after they enter; taking measures to ensure it does not lock behind them (e.g., using a spring-clamp or engaging the deadbolt) and ensuring their operation is communicated for accountability.

LOCATE AND CONFINE

The initial function of the search crew is to confirm the seat of the fire for the nozzle team. Once located, the door to the fire room should be controlled, whenever possible, which can be aided through the use of a pressurized water extinguished (a.k.a. "the can") to knockback the flames and using the reach of a hand tool to capture the door. By confining the fire to the involved area, the closed door gives the following benefits:

- Limit the fire's ability to grow and spread.
- Limit the byproducts from further contaminating the other spaces.
- Buys precious time for the unprotected crew searching.
- Buys even more time for the truly unprotected victims that may be trapped.

The same matter of access and ventilation applies for taking glass to target a specific room and conduct a window-initiated search (a.k.a. "vent-enter-[isolate]-search" or "VE[I]S" for short), as it will serve as an outlet for the byproducts of combustion, and the fire will be drawn to that new point of low pressure. Controlling the door to isolate the space from the fire and counteracting this effect, immediately on entry, is paramount. Conducting this type of search must be communicated, not only for accountability, but also, so the interior crews operating can anticipate any changes in conditions. The fire does not care why you took the window—it is bound by the laws of nature. Any time an opening that has open pathway to the area of involvement is made, regardless of intent, the fire will react accordingly— *for better or for worse*.

Photo by: *Csmeadphotography*

VENTING FOR EXTINGUISHMENT

When venting for extinguishment, the openings cannot be made until the nozzle team is *at least* on the approach to the seat of the fire (e.g., the final 10-15 feet, on average, typically starting from either the junction point of the hallway or on entering the adjacent room) and is ready to initiate their attack.

By closely sequencing the ventilation with the onset of suppression, the openings proactively created will become dedicated outlets for the byproducts of combustion to be directly exhausted, opposite/above the nozzle team's advance. The increased exchange from the pressure front being created by the air entrained from stream of the handline (5000 cfm at 150 gpm) will greatly enhance conditions within the intake path, facilitating the suppression efforts and increasing tenability. The execution of ventilation, specifically the timing aspect, will be more conservative, if any of the following factors are encountered:

- Hoarding conditions
- Deep seated fire
- Below-grade fire
- Wind-impacted fire

The environment, in these cases, can create punishing conditions and make accessing the fire challenging. Because the operational tempo of the nozzle team has the potential to be remarkably slower, in addition their progress potentially being more difficult to gauge from the exterior, additional authorization from the incident commander should be received before ventilation is conducted to prevent any adverse reactions.

VENTING FOR SEARCH

While venting for search must also be aligned with the onset of suppression, there is an alternative to ensuring it remains coordinated. For crews conventionally searching (i.e., a door-initiated search) ahead of or without the protection of a handline, they can "vent as they go" to improve the conditions in that space, as long as the affected area is (and remains) isolated from the fire first. This is essentially the reverse order of the process for performing a window-initiated (targeted) search, as described previously, but with the search being initiated from the interior door instead. By rapidly occupying the interior to conduct the primary search, these compartmentalized spaces, which are typically bedrooms (the highest probability of victim location), can be vented once the door is closed behind you after passing through it. Taking the windows within the isolated room being searched provides the necessary relief of these contaminated spaces, which would otherwise remain stagnant. Once the room is searched, the door must remain closed after exiting, if the nozzle team does not have the fire in check.

PATH OF VICTIM REMOVAL

In the event a victim is located, the vented window may serve as an alternative egress path, especially when operating on the ground-level floor or when the window leads to a substantial exterior structure (e.g., a porch, a porch roof, a balcony, or a fire escape). While it can be more time- and labor-intensive, exiting via the window can have a considerable impact on the outcome of the victim, if they are still breathing, as doing so instantly places them directly outside. If suppression has not yet begun, reopening the door to take the victim out the way you entered, on the other hand, will subject them to the contaminated hallway. Taking a victim past fire will cause their chances of survival to plummet. If window removal is not an option (based on access, the size of the victim, the available resources, or the conditions) or fire attack is or will soon be underway, consider temporarily sheltering in place, if the victim is still breathing, as your position is isolated and vented, until the nozzle team is on the approach and gains control over the fire.

SURVIVAL IMPACT

Survivability is purely a matter of time, distance (both proximity to the fire and to the floor), and shielding. While time is the immortal enemy, elevation can be the most detrimental. The degree of toxicity and thermal insult from the level of one foot to three feet in height, can be the difference between life and death, if the victim is still breathing. When a victim is discovered, immediately getting them to the floor and keeping their airway as low as possible, throughout the removal, is paramount to their survival. Being within an intake pathway and behind the protection of a closed door can further improve tenability. The power of door control to isolate spaces (the very basis of FSRI's *Close Before You Doze* public fire safety campaign) or to confine the fire to the area of involvement, as well as the coordination of ventilation openings, should be exercised whenever possible. When it comes to the avenue and method of victim removal, efficacy trumps efficiency; it does not matter how quickly you accomplished the task, if the means you employed subjected the victim to a lethal dose. If the preservation of life is truly our top priority, our tactics must be selected and implemented based, foremost, on the welfare of any trapped occupants.

Sources:
- Firefighter Rescue Survey.
- FSRI Study on Coordinated Fire Attack in Acquired Structures; Study of Residential Home Size-Up and Search and Rescue Operations.
- Leadership Under Fire: Optimizing Human Performance.

SECTION 4: ACTION

The decisions are made. The strategy has been decided. Resources have been assigned. Now it is time to get in there and get it done. The articles in this section are all about putting in the work. This is the section where the rubber-meets-the-road. The time for talking is done. It is time to take action!

CIVILIAN DRAGS
GRANT SCHWALBE

Photo by: Camera Worthy Photography

THE FULL PICTURE

The need for good primary searches has gained momentum in the last 5 years. Crews are being more intentional and training on their searches and search techniques. While this is important, we must not miss the second part of "search and rescue", removing the victim. The victim removal part of search is as, if not more important than the search itself. We have the potential to save a victim or do them more harm. If we treat our training like it's over when the victim is found, we've missed a huge piece of the training. Let's take time and dive into the topic of victim removal.

In the late 1990's, as a young firefighter, my crew was dispatched to an apartment fire with possible victim trapped. We were familiar with the location-small efficiency apartments, less than 600 square feet and all first floor with an exterior door. We ran medical calls to this location multiple times a week. While enroute my excitement was high— a fire, entrapment and a straightforward building. As we arrived on scene, we found a door wide open to the apartment and bi-directional smoke coming out. A neighbor was standing by the front door stating that they thought she was still inside. I walked up to mask-up and my thoughts were "Yes! I'm going to get a grab". But then as I started my search, my mind started to get the best of me. Once I found the victim, what was my next move going to be? At that point in my career, I hadn't been taught any drag techniques. I'd done the combat challenge, but I wasn't sure that would work if the victim was big or slippery. My brain scanned the pages in the IFSTA book, third edition for something that would work. Three firefighters holding a victim across their arms? That wasn't going to work. Two firefighters carrying a victim in a chair? Thats not realistic. And then mental block set in…all while I was searching. My quick search tempo began to slow down, because I had no clue what I was going to do if I actually found a victim. My only hope was that my partner would find the victim and know the next move, or I would rise to the occasion and figure it out. In the end, there was no victim. The occupant had left the house. But I found a big hole in my game that I needed to figure out.

PRE-LOAD OUR SOLUTIONS

- **Body Location:**
    - On the ground?
    - On furniture?

- **Body Orientation:**
    - Face up?
    - Face down?
    - Feet first?
    - Head first?

- **Size**:
    - Big?
    - Average?
    - Small?

- **Grip**:
    - Normal?
    - Use Clothing?
    - Slippery?

THE DRAGS

People often say, "Just make it happen" or "I'll figure it out when it happens". That's not how our minds work. When we are in the moment, we take in all the information and orient ourselves to what the actual problem is and then match our possible solutions to that problem. In order to be best prepared for the rescue part of the search it's important that we break down the possible patterns involved with victim removal and match our options. Training is where we "pre-load" our solutions. While we may not get a lot of sets and reps in civilian removal from fires, we do get lot on the medical calls many of us go on with similar issues.

SETTING UP THE DRAG

When dragging the civilian, we must understand a few things. First thing is that the drag is typically from a single-family house or apartment. In perspective we are looking at a 15-30' drag, not a 250' drag from the back of a Best Buy to the front door. The second thing is that the civilians are better protected if we keep them low to the ground. The traditional "Combat Challenge" drag keeps the victims airway higher up than it needs to be. Information that has come out of the Search Study by UL says, "Removal of an occupant lower in the space (1 ft above the floor) was shown to result in a lower accumulated exposure compared to higher elevations (3 ft above the floor) even if the higher elevation egress occurred at a rate that was 3 times as fast!". If this is the case, we are better off using a method that keeps the victim's airway low.

When I walk backwards a stumble on things. Having the victim in the "Combat Challenge" removal not only keeps the victim's airway up but puts me in a position to be walking backwards. If I stumble and fall there is a chance that the victim ends up on top of me, which is bad. There have been several accounts of this happening in the interviews I have done for the Grabs Podcast. One good part about the low drag is that if you need to adjust or take a break you do not need to set the victim down and then pick them back up. They are already in a low spot.

BODY LOCATION

We must first get the body to a place where we can drag from. If the victim is on a bed, couch or chair we must get them to the floor. On a bed we can use the bedding material to make a "burrito" and lower them to the floor. If there are no sheets or blankets underneath them or they are on a couch or chair, it's easiest to grab under the shoulders or under the arms and move the victim to the floor, while protecting the head.

The drags all work easier if the victim is on his or her back. Having them on their back also protects the face. If you find a victim face down, the easiest

SECTION 4: ACTION

way to get them on their back is to take both of the victim's arms and place them over their head in line with rest of the body and cross their arms. Take the opposite arm of the side you want them to flip to and place it on top of the other. When you pull the body naturally flips to the back. If they are large or the body doesn't flip easily a second rescuer can help turn them at the hips. Once the patient is on their back, they are ready to drag. Victims present either feet first or headfirst. My preference is the feet first drag if possible. This may require you to spin them around. If there is not enough room to spin them, you may do a headfirst drag to move the victim out of a tight place first and then spin them. To spin the victim, simply take their feet off the ground and spin the body on their back (like the old school breakdancing move).

HEAD-FIRST ARM LOCK

Bring victim's arms above their head and cross their arms at the forearm or wrists. The rescuer reaches through the arms (under) with his/her dominant arm and then locks the other arm on top (sandwiching the victim's arms together). If you try to "grab" your hands or use your grip it will fail very quickly. Instead try to visualize a trailer hitch. Your bottom (Dominant) arm is the hitch, and the top hand keeps it together. As you begin to drag the victim's arms come together and bite on yours. This is a great drag for 10-15 feet. If you try to practice this using no gloves, folks often resort to trying to grip the victim's arms and it fails, so be sure to practice this using fire gloves.

HEAD-FIRST SHEET/WEBBING DRAG

If the victim is slippery or a large victim, it may be necessary to use something to help get a grip. You can use a sheet, drapes or even pants-or any other items you can find in a room to help with the drag. Place whatever you are using at the victim's head. Either sit the victim up or use the arm drag to get it behind the victim's shoulders. Bring the ends of the sheet (or whatever you are using) underneath the armpits and up to the head. Webbing can also be used in the same manner. You may need to double up the webbing to shorten it to the right length.). This method easily allows for a second rescuer to assist in the drag if necessary for a large victim.

FEET-FIRST LEG DRAGS

This drag uses the victim's legs as handles. The rescuer reaches on the outside of the victim's legs getting the elbows behind the knees. The rescuer squeezes into the body. The torso-the heaviest portion of the victim, stays on the ground. The key in this move is locking your elbows behind their knees, joint to joint. While the rescuer is moving backwards, they are often led by another firefighter, and they are moving from an area they just came from. In this drag the rescuer is up high.

Should the rescuer need to be low, another feet first drag may be preferred. In this drag the firefighter has their front knee and front hand down. The back leg is up (Nozzle Forward Heel Position). With the victim on the ground put one of the victim's legs over the other. They will be crossed at the ankles. With your back arm, grab the victim's legs and keep them tight to your body with the ankles binding in your elbow joint.

FEET FIRST SLIPPERY VICTIM

If the victim is burned and difficult to hold on to you may need to try the Feet First Girth Hitch. For this drag I use a 7' prussic (8mm rope) tied in a loop (double fisherman's knot). To secure the victim, place the loop over the victim's feet and girth-hitch it around the feet. This drag also allows two firefighters, if needed, to drag the victim, especially if dealing with a larger victim. Be mindful when dragging the victim in this method, that you have lengthened the "wheelbase" of the victim and turning corners may be difficult. You may need a firefighter to act as the "tillerman" to get the victim around corners.

When doing drags, body positioning is important. When pulling the victim, it's quickest to move in a tripod position. The most efficient stance is front hand down and front knee down, back leg up. This allows you to squat the victim's weight rather than row it. Consider the statement "Heal is the motor" from Aaron Fields (Nozzle Forward) when getting into position.

THE WAY OUT: FIRST DO NO HARM

Put yourself in this scenario: You are assigned primary search. You and your partner crawl through a dark, hot, and smokey house. You pass the engine company putting water on a kitchen fire. You and your partner make it to the back bedroom and find a closed door. You open the door and find light smoke conditions. While searching, you find a 50-year-old, 135-pound woman unconscious on the bed. What's your move?

CREW-DISCUSSION

- Do you take her out the window because it's the closest point to you?

- Is it quicker to go back where you came in from?

- What happens if you pass a room on fire?

Let's take the same scenario but instead the bedroom door was open and the conditions in the room were hot and smoky, just like where you came from.

> **CREW-DISCUSSION**
>
> In the excitement, do you take her out the way you came through… the hot, smoky environment? If so, you did more harm than good. She was in a safe spot. Remember- "Closed doors save lives"
>
> - Do you choose window removal?
> - Have you practiced window lifts?
> - Is this the first or second floor?
> - Is a ladder thrown to the window already?
> - How long will it take to break the window?
> - Not just break the window but clear it enough so that you are willing to hand out your mom through it?
> - Does this scenario change if she is 350 pounds?

TIME, DISTANCE, & SHIELDING

"Time, Distance and Shielding" is often used in the Haz Mat world, but I like to apply it to the victim removal world. It's simple and common sense.

> **TIME, DISTANCE & SHIELDING**
>
> - Limit the victim's **time** in smoke
> - Consider **distance** to the closest/fastest exit
> - **Shield** the victim from heat and smoke as much as possible on the way out

More hands does not always make for an easier drag. Often when more people try to help, they get in the way. Have a plan for the others that come and want to help. The first one is to have someone lead you out. It's much easier to have someone navigate your path out by grabbing your pack and pulling you in the right direction. They can also kick stuff out of your way that may impede your drag. If you have additional help, use someone specifically to clear the path. You may end up needing one or both of these members should the clutter prohibit a drag and make it necessary to carry the victim over debris. If there are additional fire-fighters in the house or vicinity, doing non-essential tasks, you may consider passing off the victim to them for removal, so that you can finish the search. You are already familiar with where you've searched and can likely complete it much quicker. Obviously, if there are no free members to pass the victim off to then do a victim removal; but do it because it makes sense for the victim and not because of your ego.

> **PRE-PLAN YOUR TACTICS**
>
> - Find what options you like for headfirst, feet first and slippery/large victims
> - Identify the scenarios when you will use each option
> - In scenarios, make the opportunity for recognizing the position of the victim and matching the appropriate drag based on the victim and conditions

We must continue to train on search. Part of that training must include the rescue or removal of victims. The training should occur in 3 phases.

Having a plan and figuring out what works for you. Simple, quick options that are repeatable under stress are essential. That, matched with good search techniques will assure you are prepared for your next structure fire with victim trapped

CONSIDERATIONS FOR VICTIM REMOVAL

DAVID (DJ) STONE

Photo by: Camera Worthy Photography

CONSIDERATION #1: TOOLS

Before the victim removal is accomplished the determination of tools needed must be defined. It is recommended that tools required are limited. During a search the most important tools you have are the ones you are born with. Your sense of sight, hearing, hands, and feet. If you can see it, it is searched. No tools on an active search allow us to use our hands and feet to sweep, clear, and move quickly. Targeting searches visually significantly speeds up the process of discovery. The other tools considered to carry with the search team must be easy to carry and use.

It is suggested that the can and a set of irons are the tools of choice for a search team. The can provides the search crew with protection for the team, although it can slow down the search. A tip is to stage the can in a strategic location when performing a search and remember once the can has been expelled it can be left behind if necessary.

Not carrying a tool as the lead target searcher allows the member to be free from subconscious management of a tool in hand. Leaving the tool with the oriented firefighter frees up the searcher to move about more rapidly. An always rule should be that the oriented search partner (holding tools at the point of entry) be within verbal or visual contact. At a minimum, the tools can be left at a doorway or point of entry as an orientation. When we encounter a victim, we will be prepared to immediately assist without a tool hampering our progress.

CONSIDERATION #2: THE ENVIRONMENT

What are the conditions found upon entering the environment to be searched? Size up is everyone's job on the fireground. The ongoing survey of the scene and immediate operating environment is critical to the safety of the search. We must take into consideration the environment encountered inside "the box," aka the area or room of search, and its effects on the removal of the victim. If conditions are not viable for us, it is not viable for them thus affecting the removal process. We should consider door control as we enter a space to be searched. As minimum staffed suburban companies, sometimes the engine is simultaneously searching for fire and victims. If victims are on that path, they get removal first. However, we must be cognoscente of controlling the environment with a dedicated nozzle position or closing a door.

During our size up and upon the approach we must consider the following concerning the smoke conditions; what are the smoke conditions telling you?

Visually, is it angry fast smoke? The door is open, and the fire is near. Is it slow and lazy? The door may be closed, or the fire is distant. Either way speed is necessary for the viability of victims.

Simply put, the hotter the more volatile the environment, the harder and riskier the removal process will be and more stressful for the victim. This means our time is limited unless we control our atmosphere. The two ways we can control our atmosphere are closing the door or applying water. The situation will dictate which tactic is appropriate. When inside the room understanding the environmental indicators will help us determine the time, we have to remove the victim. The ways we can monitor our environment are with our senses

Photo by: Camera Worthy Photography

(eyes, ears, etc.) and TIC's (thermal imager cameras).

The TIC: convective currents indicate the path of fire. Scan high, when currents are present this means there is an open path to your location. You may need to close a door or apply water. Time is limited. Visualize the environment through the eyes of your thermal imaging tools. Act fast to control that box.

The environment for the victim is obviously dangerous and toxic, for us it is routine business. Our gear affords us the ability to take the assault when encountering conditions inside the building. We must understand when it is time to abandon the search for victims and when their survivability is in question. Keep it simple; if it is hot enough that your shoulders and upper body are becoming notably hotter (in your gear) when in the crouching/crawling position, it is quickly becoming untenable inside.

Two things need to happen if the box is not controlled. Get out (abandon the victim) or flow water. If the box is controlled by the door, you have time. Depending on the environment we encounter it will affect our decision to remove the victim. Situations will dictate the environmental conditions encountered on arrival. Regardless, you will have extremely limited time to make that decision and determine the risk. When we can isolate the fire area, it allows a search to continue. Otherwise, we are battling the clock, and unchecked fire growth.

Is the environment hoarded or severely cluttered? Most likely it will be. This remains the most unpredictable consideration for removing victims. Simply put, the more stuff in the way, the more difficult the removal will be. This requires resources and calling for them quickly. However, in minimum staffed situations it may only be two firefighters to affect a rescue. In this case it will require work to clear the path and determine the quickest route to safety and clean air.

Photo by: Camera Worthy Photography

CONSIDERATION #3: QUICKEST ROUTE OF EXIT

What is the route to find the victim? Depending on the conditions encountered which way is the best way out? The majority of the time it will be the way you came into space, however; we cannot predict when or where we will move our victims. Without some sort of fire control the search is always endangered. Having an oriented partner is key to direction and alternate routes. The team must remain oriented and aware of our surroundings and the options they present. Typically, that person will have mapped the way in and hopefully considered the building prior to entry and assist in determining the way out.

Keep it simple; if it is a straight shot back from entry (fire mostly controlled) and path is clear, it may be best to head out the way you came in. In other more volatile circumstances where you do not have the protection of a hand line, it may require you to move the victim to an area of refuge within the structure until you can get assistance or better control the environment.

In any case, you need to have a plan to move the victim outside of the structure.

CONSIDERATION #4: VICTIM SIZE AND REMOVAL

This is the most critical factor in victim removal. Is the victim large or small? Clothed or not? Is the victim oriented to the direction out? Can we get them out with 1, 2, or 3 people? What resources will I need to remove an extra-large victim? We will have zero intel and control on this matter.

We must have a plan, and that plan must be simple. Keep the moves to 3 steps or less and get assistance. The most crucial factor in victim removal is keeping the victims' head low to the ground. Recent FSRI studies have shown that the cleanest air inside a structure fire is closest to the ground.

Considering that information, the most efficient way to move victims is with the thigh drag; grasping the victims' legs high up at the thighs, locking the legs into your arms, and dragging the victim to the desired location. In doing this maneuver you are utilizing your core and legs to move the victim, while keeping their head as low as possible to the ground. It is recommended that any time you encounter an adult victim of any size, to get assistance from your partner. Moving victims in real life situations in hoarded environments is extremely difficult and is made easier with assistance. Your partner can assist in the movement of the victim when standing is not an option.

Another way your partner assists is communicating with command victim contact and removal. Large victims will require assistance especially when removing victims from windows. Resources may be needed.

Depending on the size and complexity of the victim's location, this may require some form of area of refuge. Extra-large victims may not be able to be moved over long distances and may require moving to a room and closing the door. In this case when the victim is so large that they are unable to be lifted to a window for removal, a window conversion (Cutting away the window into a door from the base down) may be in order. Again, situation will dictate such extreme options.

Finally, what is the plan when the victim is removed to clean air? We must commit to patient care and initiate EMS (Emergency Medical Services) measures as soon as possible. If patient care is being rendered, get back inside, and finish the search. You are their only hope.

CONSIDERATION #5: TRAINING & PERFORMANCE

We will perform in the way we practice. Training in how we make decisions and move victims is key to success. Keeping training simple and highly repetitive is crux.

Searching in structure fires is considered for most a low frequency, high risk event. Considering the dangers that present themselves in these types of events, we must practice to a point of minimal thought to engage in action. In other words, we must drill to a point where we default to a specific body language, communications, and victim movement. Hands on training and building a mental muscle memory creates an ability to think more rapidly and move to action.

> *"When you train on the fundamentals... drilling on them, it frees your brain to think through the variable. To recognize when you need to call an audible or change tactics."*
> ~ Grant Schwalbe

Rapid decisions concerning victim removal will be based on your experience, and this will mostly be built on your training and what you feed your mind. As we gain experiences in real life, we will then be able to integrate those incidents into our training— forming a more dependable course of action.

Train in real time, with real people, in hoarded type of environments for best results.

The keys to success in the fire service are keeping operations as simple as possible. The less we process, the quicker we act. Our safety in operations is built on the knowledge you put into your mind and the practice you put into the field. Add experiences in real life (which take time) and success will certainly follow you.

WINDOW BASED RESCUE
MATT DONEY

Photo by: *Csmeadphotography*

WINDOW WORK

As firefighters we wouldn't think that something as simple as taking a window could be taxing. Believe me when I say they can very easily humble any firefighter that doesn't have an understanding what this really looks like on the fireground.

What will help is having an understanding of your enemy. What type of window are you up against? Wood? Metal? Vinyl? Single pane? Dual pane?

Once we have sized up our window (this should take seconds) and are ready to take glass, we want to make sure we are prepared for immediate entry by having our mask on before we take the glass. Once you swing at the glass you have started the clock and seconds count. Victims on the inside do not have time for us not to be ready to enter immediately.

The Halligan is a great tool to take windows with. It weighs just enough and a 30" bar allows you to choke up to put the full head weight into the sash. The glass is the easy part, aim for the top, once you break it let it burp for a second to see what's going to meet you outside before making entry. Don't be robotic, we have training scars when it comes to taking sashes. If the sash left in place allows access and egress leave it in place. If you have to take the sash, do it as quickly as possible.

Wood sash is generally no problem for us. It typically breaks away easily and does not take much time and energy.

Metal sash can be a bit more problematic if the window is well built. However, more often than not the metal will fatigue and break away after a couple well placed swings.

Figures 3 and 4: Metal frame, single pane

Figures 1 and 2: wood frame, single pane

SECTION 4: ACTION

Vinyl windows are a different animal. They typically require more time and energy to clear. I have had 8lb tools bounce off them like a trampoline.

Figures 5: Vinyl frame, dual pane

For these types of windows, take the operational side or section of the window (meaning the section that moves) out of the equation prior to clearing the sash. After taking the glass, put a good swing on the sash and the operational section will distort from the fixed section. This allows you to easily pull it out or push it in off it's track with a tool. Now you're only fighting with the smaller, weaker portion of the window. Just remember, if the window is big enough and the sash will allow for access and egress then leave it in place. Clear the glass and go. If working from a ground ladder, be sure to soft lock or at minimum push your legs out wide beam to beam.

Figures 6 and 7: With vinyl windows, swing against the direction of the seam to more easily defeat the sash

Page | 193

CURTAINS AND BLINDS

Be careful using tools to remove curtains and blinds. The old 'twisting your hook in curtains' trick is great until you're on the fireground trying to untangle them from your hook. Similar challenges arise when dealing with mini blinds. These should be removed by pushing the header off its track or hooks.

For removing curtains, twisting the head of your tool works, however, now you have to untwist the curtain off the tool with your hands. Instead, removing the curtains only using your hands is the easiest, most efficient way.

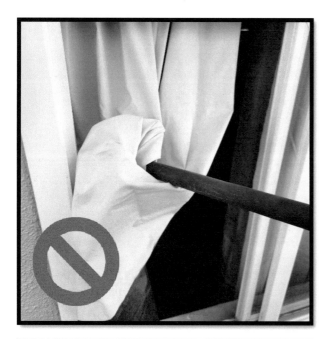

Figure 8 and 9: Removing curtains

Figures 10 and 11: Removing blinds

Do not put your hook or halligan in the blinds for removal. They will become tangled and take some effort to remove. Push the rail inside and drop the entire blind assembly.

When your window is clear for entry it's time to get inside. If the floor is visible from the outside, I'll straddle the sill and step inside. Once inside you should still get as low as possible to gather an immediate Life, Fire, Layout. If the window is dirty and you can't see the floor you should go in headfirst. This method has its advantages. It will be easier to overcome obstructions on the way in, like dressers, nightstands, etc. Additionally, by entering headfirst you, by default, search the first 6' to 8' inside the room immediately. Now you're also in the most advantageous position to perform your Life, Fire, Layout and locate doors for isolation if needed.

VICTIMS PRESENTING

A common problem we see both on the training ground and on the fireground, are firefighters that are given an objective and become so mission focused on that one thing that they are moving around the building without having their eyes on it. We can miss a lot of important things. Victims hanging out of windows, which windows are clean, and which are dirty, curtains or blinds moving from a civilian inside. We want to quickly identify where our victims are or where they might be and we can't do that if we don't have eyes on the building.

Figure 12: Here the firefighter is moving around the building with a heads-up approach.

You and your company should train and have a plan for conscious victim rescue. If this is something you haven't thought about or trained on, it could go very slow. Vital information can be missed, and we could be delayed in reaching other potential victims inside.

STEP ONE:
VISUAL CONTACT

If we are coming around the building and have a victim presenting, we want to make and keep visual contact with them. What you don't want to do is turn your back to them in an attempt to retrieve a ladder. Our objective is to keep them at the window and make a radio transmission for a ladder if you didn't bring one with you. If we leave them, they may suck back into the building, take a couple hits of smoke and go down. Then we will have a bigger problem that requires more time and energy to mitigate.

STEP TWO:
VERBAL CONTACT

Next step is to make verbal contact. This is how we keep them as calm as possible. Reassure them we are coming up to get them. If they look like they might jump prompt them to "stay there" or "hold on" and start asking questions. Is there anyone else inside? How old are they? Where are they located in the building? This is important information to relay to your partner and/or command. This information should prompt crews to occupy and search those areas immediately.

STEP THREE:
PHYSICAL CONTACT

Once your primary ladder is up, the next step is physical contact. I will not even take the time to put my mask on, if the victim can deal with it so can I. By having my mask hanging, I have the ability to share my mask with the victim if needed. Climb the ladder, grab a hold of the victim, and don't let go for the duration of the ladder rescue.

Now the coaching process begins. These victims may shoot out on top of you before you're ready for them or they can be moving painfully slow. Keep in mind this civilian may have never been on a ladder before or could be afraid of heights. Add the stress of a working fire on top of all this and they may not even hear your commands. Make sure to be loud and firm in your commands, make it clear that you're in charge and you are helping them. Listen to your victim, what are they telling you? If they aren't telling you anything and you haven't already asked the pertinent questions mentioned above, then now is the time to ask. "Is there anyone else inside? How old are they? Where are they located inside the building?" This is all important information for you to know and to relay to your company or command.

Figures 14 and 15: Here the second firefighter has thrown the secondary ladder within stepping distance of the primary ladder that he will use as a passing lane to get inside (top photo). The second firefighter has thrown the secondary ladder and is masking up for window entry (bottom photo).

Figure 13: Get hands on your victim immediately. Get their airway out of the exhaust. You can use the building to protect their airway by bringing your victim's head out to the side, keeping it as low as possible.

As this is taking place your partner should be throwing a second ladder within stepping distance of the primary ladder. This allows your partner to use it as a passing lane to get inside fast especially if your victim is reporting others inside. It also allows for a second set of hands at the window if needed. Have the mindset that anytime we have a victim conscious or unconscious, we want double ladders. This will also be beneficial to us if our victim is a large unconscious adult, and we need to perform a window door conversion.

SECTION 4: ACTION

Figure 16: Here we see the second firefighter stepping over to the primary ladder for entry. He has bypassed the slow process of the rescue.

Figure 17: Here the firefighter has coached and assisted his victim out the window. He is not allowing too much separation between himself and his victim. He is in a good position to pin his victim if they go unconscious or start to fall.

Coaching your victim out of the window should look the same way we would come out of the window. Have your victim hold onto the window frame as they bring one leg out to straddle the sill. Make sure they are straddling the sill with their head outside and place their outside foot on a high rung. Have them step up and bring their other leg out as you guide it onto the same rung. Again, this looks no different than the way we transition out a window to a ground ladder. Once on the ladder make sure you don't allow too much separation between you and your victim. Stay close, your head should be at their hip level just in case your victim goes unconscious or starts to fall you're still able to pin them into your ladder and call for help.

Figure 18: 2nd firefighter moving past the first and continuing the search.

At this point in your rescue your partner should be masking up and coming up the second ladder. This is when you relay the information your victim told you. Additional victims, ages, possible locations. Your partner will step over to the primary ladder for window entry.

VICTIM PLUS VICTIM

Let's take a step back now and say not only do you have a victim presenting but they are holding a baby or child out the window. This should change the roles of the firefighters. The first firefighter should drop everything and immediately put themselves into a position to catch the child if dropped.

Figures 20: Here the firefighter sees a child being held out the window. He is still moving toward the window while dropping everything he is carrying in anticipation of the child being dropped. The firefighter has put himself in a position to catch the child if dropped. He is prompting his victim to hang on and do not drop the child.

Figure 19: Victim presenting with a child.

Don't encourage the victim to drop the child. Encourage them to hold on to the child as your partner throws the ladder and gets to them as fast as they can. Once we make it to them grab them both and call for your partner to get the child from you. This way the victim is not able to go back in the room for another family member or a pet. Once you have passed the child, coach your victim out and continue your normal window-based search operations.

Figure 21: A second firefighter has thrown the ladder and has a hold of both victims

SECTION 4: ACTION

When providing patient care to your victims, make sure to bring them far enough away from the building and provide care facing the building. We can remain a fire ground asset if we are able to keep eyes on changing conditions for our crews operating on the interior.

Figure 23: Here the firefighter has remained an asset by distancing himself and his victim from the building. He is facing the building and is able to keep an eye on changing conditions.

Figure 22: and takes the child down to safety. This allows the firefighter still on the ladder to start the process of getting his victim out of the window.

SEARCHING IN HOARDER OCCUPANCIES

TIM KLETT

When we take an in-depth look at fire ground operations, there are very few actions that are more challenging and dangerous than the primary search of the fire occupancy. Notwithstanding the challenging nature of these operations, it is truly the obligation of every fire agency to attempt to locate and remove trapped civilians, when and wherever possible. These dangers can and will be multiplied depending on the occupancy, building type, fire condition and most importantly the condition of the area to be searched.

Certain occupancies will require a specialized search such as a commercial building, where most experts advocate the use of the search rope for a large area search. Normally, we can train for these specialized searches to perfect the use of the tools carried to conduct such searches. Unfortunately, the condition of the area being searched is a problem which carries so many variables it can become difficult and time consuming to perfect your search techniques for every possibility.

IDENTIFICATION IS KEY

One such situation which poses a serious threat to civilians and firefighters alike is an occupancy with a medium to heavy clutter condition, better known and often referred to as a

hoarder house. One of the most important actions of the first arriving officer is to identify hoarding conditions and to locate the point which the occupant is using as their primary access. At some hoarder houses the property will mirror the occupancy, making it easy to identify and at others the property can be pristine. There are many obvious and subtle clues that can tip our hand into believing we are dealing with a hoarding situation, but prior knowledge or occupant information are the most reliable ways.

When trying to determine the point at which the occupant enters and exits the house there are some distinct indicators. We can look for paths through exterior junk and garbage, paths worn in grass or snow, cars parked deeper down long driveways and shopping carts, wagons or carriages stored at a doorway. We must be aware that these entrances will vary based on the occupancy involved, whereas a private house can have multiple access and egress options while a multiple dwelling normally has only one. It is also extremely important to remember that most interior actions will be initiated at this doorway, whether or not it's the normal primary means of egress. In addition, once this access point is located it must be clearly defined too all members operating keeping in mind that due to the hoarding conditions, we should be more concerned about our ability to get out of the building than we are about getting in. Remembering that this fact does not relieve us of the responsibility to conduct aggressive interior operations, it simply means that we must always be vigilant in clearing pathway to our way out.

INTERIOR OPERATIONS

Regardless of the occupancy, once the primary access point has been established companies can now focus on beginning interior operations which includes but is not limited to the primary search. Arriving units must understand that most hoarder fires will be vent limited and haphazard ventilation can have serious consequences within the fire occupancy, endangering civilians and firefighters alike. Unlike searches conducted under normal conditions in which firefighters should penetrate to the seat of the fire and search their way back, searches in hoarder occupancies should begin immediately upon entering the occupancy, this is true in both private homes and multiple dwellings. The interior search team must understand the emphasis of allowing the engine company access to the fire area within the occupancy. The hinge pin for a successful search, during hoarding conditions, is quick water on the fire, which allows for a more aggressive ventilation of the occupancy. As stated before, ventilation must be managed and timed more closely at these type fires because our interior operational time values can and will be elongated.

With any primary search size-up, it is imperative to quickly identify the areas that require immediate action based on the location of the fire and the access points to initiate the search. When searching hoarder occupancies interior crews need to understand that some of our normal search practices cannot always be applied. Occupants can be found in almost any area or room at any given time of the day or night. Simply put, it has been found that during hoarder searches, the junk ends up where the people should be and sometimes the people end up where the junk should be. This epiphany of information translates into the fact that bathrooms and closets are required to be searched the same as bedrooms and kitchens.

CONDITIONS DICTATE OPTIONS

Once the search crew enters the building it must determine the type and severity of the hoarding conditions, which can dictate our search strategies. With most medium clutter conditions, without pathways, an up and over approach can be successful in achieving our search goals. It must also be emphasized that while searching, any object that poses an entanglement threat must be removed before continuing. When dealing with pathways, which are very common with hoarding in multiple dwellings, the search team must make an evaluation of the viability of the path itself. Items stacked high along the path can collapse on or behind the search team seriously endangering the members. If the path is deemed too hazardous to use, an alternate route must be found to access the areas needed to be searched, in multiple dwellings this might require breaching from the adjoining apartment. If interior crews encounter a heavy clutter condition the primary search can be seriously delayed due to the difficulties of navigating the junk itself, this might require the removal of substantial amounts of items. It should be stressed that when moving trash, the areas used to put the items removed must be searched prior to filling them. Another method that can be of assistance when operating with heavy clutter is to use items such as doors, headboards, or anything solid and flat to build a floor on top of the clutter. This will allow search teams to access areas quicker and aid with the removal in the event a victim is located.

VIGILANCE IS NECCESARY

Inside teams must be hyper vigilant of their surroundings due to the ease of becoming

disorientated, the clutter can become a giant quagmire making areas undistinguishable from one another. When dealing with severe hoarding inside teams can identify a large item, such as tires, desks etc., to use as a rallying point in the event you get turned around or conditions change rapidly. It is always a good practice to leave a member on our avenue of egress to function as a beacon to other members of the search team, regardless of the severity of the hoarding. Crews conducting searches in hoarder occupancies will undoubtably be lifted off the floor due to large amounts of clutter. During the search members must constantly be aware that when moving from room to room or hallway to room they are passing through a door frame and not a window frame, remember that once we are lifted off the floor our perspective of the area becomes distorted. Due to the varying levels of junk search crews should always try to maintain visual contact with their partner to minimize the chances of becoming lost or disorientated. At normal fire operations smoke will always distort Time, Distance and Sound, at hoarder fires, with the addition of copious amounts of trash, this distortion of time, distance and sound will be magnified dramatically.

THERMAL IMAGERS

The use of a TIC (thermal imaging camera) if available, is highly effective during normal primary searches but is invaluable during the search of areas with hoarder conditions. The TIC allows the search teams to quickly scan remote, impassable, or even unreachable areas for missing or trapped occupants instead of wasting time trying to access those areas through heavy clutter. If the TIC shows any distinguishable heat signature representing the possible location of a victim those areas must be entered and quickly searched. One drawback to the use of the TIC is the actual landscape within hoarder occupancies where severe hoarding creates hills and valleys. In this situation the search team must be cognizant of the fact that they may have to climb to the top of the trash to scan an area looking downward.

VES CONSIDERATIONS

When manpower and conditions permit, a two-pronged approach to the search can be very effective in locating trapped occupants remote from our primary point of entrance of any respective occupancy. Accessing the building via windows is a tried and tested maneuver that has proven to be highly successful in rescuing occupants. The use of VES (Vent – Enter – Search) tactics at hoarder house fires can be very beneficial to the search but, it is not as clear cut as it is when dealing with regular structural fires and some controls will need to be attached. Once again, as stated prior, hoarder fires are vent limited fires and uncontrolled ventilation can have adverse effects within the occupancy. VES actions should never be attempted arbitrarily at hoarder fires unless there is a known life hazard in the room the VES is being performed. The ability to isolate the area being searched, by closing a door, will always come into question in occupancies with hoarding conditions. If the member performing VES cannot successfully isolate the room, they have only succeeded in venting an area that is probably not ready to be vented which can and will lead to further delays of the critical primary search. When crews are considering the use of VES for searches at hoarder fires there is an important caveat attached, they must obtain permission from the incident commander or, in lieu of the IC, notify the 1st. due engine Officer. The reason for these notifications prior to the onset of VES is to ascertain the location of the nozzle team in relation to the suspected fire area. If the engine is considered close enough to the fire area to intervene in the event of a rapid change in conditions, then all VES should be allowed. But if the engine is struggling and has not been able to locate or reach the fire area then a delay in VES activities must be considered.

STAY FOCUSED

Fires in hoarder houses can be extremely challenging and hazardous, search crews must be focused on their ability to safely penetrate to all viable areas within an occupancy while also dealing with deplorable conditions. Through training and awareness fire departments can and have had successful search missions within hoarder houses. The key to these successful missions is the close coordination between the Engine and Search crews with special attention placed on the absolute control of the coordinated ventilation of the fire occupancy in relation to the engines ability to get water on the fire.

Search crews must remember that hoarder conditions will flip the fire ground in the fire's favor, and they must modify traditional search tactics in an effort to conduct an effective primary search. Once the engine company is able to flow water, coordinated ventilation can begin, which will allow search crews to apply normal primary search tactics. It is the absolute obligation of any fire department to conduct rapid primary searches, regardless of the conditions. With training, situational awareness and determined firefighters these challenges can be overcome, which will lead to successful search missions.

WINDOW CONVERSIONS
MATT DONEY

WINDOW CONVERSIONS

The American fire service tends to hyper focus on specific events that lead to us performing annual drills such as the Denver. Not to discount this drill, its great and has value, however, what we don't do is train our membership on any other alternatives such as the window to door conversion. Here we will take a dive into the details of this operation and with proper training and equipment, companies can have this done in 60 seconds or less when removing a large victim or a down firefighter is necessary. We will run through what this looks like while performing this conversion from a floor 2 window. The ground floor is the same operation minus the ground ladders.

To be successful, we must be able to convert these windows quickly. Let's first take a look at building construction. If we are looking at the picture above, we will see the sill highlighted in green. We have jack studs that are highlighted in yellow and typically a center support that's highlighted here in red.

First things first, we want to get our ladders in a position that allows us to get a good angle with the saw, so we are not reaching or attempting to cut off balance or at a weird angle. We also need to get a good look inside to ensure it's clear on the back side of the window we are converting. Ladders are best positioned at sill level on the outside of the windows. This allows a comfortable saw reach and allows you to get eyes inside prior to making your cuts. Once your ladders are in position, FF1 heads up the ladder with a hook and a saw and FF2 heads up the ladder with a hook.

Figure 3(above) and Figure 4 (below): Ladder placement to begin the window to door conversion. Ladders are in place, FF1 heads up with a hook and a saw, FF2 heads up with a hook.

Once we are at the tips and ready to cut, FF2 must ensure the room inside is clear of any furniture, clothing, drapes, blinds or anything that will potentially put the saw out of service if the bar comes in contact with it. Something as small as a toddler sock will put the saw out of service and slow down the operation. FF2 reaches in with his hook and clears the floor for the saw and makes sure the inside team is clear of the saw work about to take place

Figure 5 (above) and 6 (below): FF2 leans in to get a good look, clears the floor, and ensures the rescuers and victim are a safe distance from the window.

SECTION 4: ACTION

Now we are ready to make our cuts. This is where the building construction understanding comes into play. Remember the jack studs? These are what we want to avoid. You are not going to cut at the width of the actual window. You are going to move the bar in about 4" to 6" to make your cuts in an effort to avoid plunging into the jack studs. Avoiding the jack studs saves time, energy, powerhead and chain. FF1 hangs his hook, assures the inside team is clear once more and begins his cut. Once FF1's cut is complete he will pass the saw and transition to his hook to control the work as FF2 completes the conversion. Once the cut is complete, we want to pull the work outside the building. FF1 who is on the hook waits for the saw to tell him to pull the work out. If we get too hasty on the hook, we can pull the work right into the sawyer causing him to fall or drop the saw. There could be times where a 3rd cut will need to be made horizontally at the bottom to pull the work out, this is typical for balloon framed construction.

Some companies choose to cut away the carpet below the window prior to making the cuts. This can take additional time depending on the type. Carpet is 50/50 when it comes to putting the saw out of service. One way to avoid this is to not cut all the way to the floor, stop short of the floor and muscle out the rest. FF2 can drop his hook inside the window to the floor, slide his hand to the sill and pull it to the outside to show the sawyer as a point of reference.

Figure 7 (above) figure 8 (top right), and figure 9 (bottom right): FF1 completes initial cut. FF1 hands the saw over to FF2 and transitions to his hook to control the work. FF2 completes his cut. Once FF2 is done, he clears out of the way and gives FF1 a thumbs up to pull it.

Figure 10 (above) Here FF2 is measuring the sill height by dropping the chisel end of the hook inside the window to the floor. He then slides his hand to the sill. Now he can show the sawyer where the floor level is so the saw can be stopped short of it. Figure 11 (below) The hook is held outside the window for reference. Here FF2 shows the sawyer the distance to the floor using his hook for reference.

When the saw work begins, keep your bar depth in mind. The more bar you have inside the windowsill, the more trouble you can run into jamming up your saw.

Figure 12 (above) Here the firefighter has the bar too deep into the building. It's easy to jam up your saw with too much bar inside the window. Figure 13 (below) illustrates proper saw depth. It's just enough to get through the material and stay clear of interior debris and rescuers.

SECTION 4: ACTION

Figure 15 (above) Firefighters reposition the ladder tips to the bottom of the sill to receive the victim. Figure 16 (below) Finished product from the tips.

Figure 14 (above): The work is pulled out rather than being pushed in where it can hinder or clog up your victim removal. The saw is knocked out of the sawyer's hand by a hasty hook. Wait for your sawyer to clear the way prior to pulling the work out.

Now that your cut is complete, and the work has been pulled out, it's time to relocate your ladders to a position to assist in removing the victim from the conversion. Ground ladders should be repositioned to the tips below the conversion, so the tips are just below floor level. You and your partner ascend the ladders to the tips and receive the victim from the inside team. The pictures below are some examples of what this operation looks like once completed.

Figure 17 (above) - Finished conversion from the inside perspective.

This conversion is often thought of as a rescue tool, however, it has a number of uses on the fire ground. My organization will use this for something as simple as overhaul. We try to avoid injury by converting the fire rooms window and carrying things out versus attempting to throw it all out the window. It gives your company an opportunity for a rep and you're not carrying things through the house to get them outside to overhaul. We will also apply this tactic to hoarder homes. Once we have anyone operating on the interior of a hoarder home the Truck will pick some key windows to convert to give the interior companies additional means of egress. Basement fires are also another application. Windows are converted to doors then the floor is opened up just inside the window conversion. Again, this allows for multiple points of attack and or multiple points of access and egress in an area with normally limited access.

Good luck and have fun.

CAN CONFIDENCE:
LOCATE, COORDINATE, ISOLATE

ERIC WHEATON

Civilians trapped by fire have a greater chance of survival when search and suppression are executed simultaneously. To add further value, the primary search team, working ahead of the first in attack line, has the opportunity to locate, coordinate, and isolate when searching with the 2 ½ gallon water extinguisher, also known as the water can. The water can is arguably the most underutilized tool found on nearly every fire apparatus in the country. It allows well-trained firefighters to quickly place water between fire and potential victims. This quick water, coupled with a closed door, is the optimal addition to the primary search. Now, let's dissect searching ahead of the hoseline with locate, coordinate and isolate.

LOCATE

The first step into gaining can confidence is locating the actual fire. This is the most important step when conducting search operations ahead of the hoseline. We must locate the fire quickly to take full advantage of the water can. Now, do we search to the fire, or do we move straight to the fire, bypassing searchable areas, and search back? These are two questions firefighters frequently ask. To answer these questions, we must ask ourselves, is the fire location known or is it unknown?

If the exact fire location is questionable or unknown, then the primary search team must focus their search towards the area that best indicates a fire's location; sometimes this is a simple guess. Being knowledgeable in building construction and fire dynamics will help facilitate a clearer read of the

fire building and its conditions, likely guiding you to the fire. However, searching towards the bedrooms is a good rule of thumb if the fire location is unknown. According to the firefighter rescue survey, this will put you in the area most likely to find a victim. And, if you happen to find fire in and around the bedrooms, then you have put yourself in a place where unprotected victims are most endangered and in need of imminent rescue.

If the fire location is known, then the primary search team must focus their path of travel towards the fire's location. Remember, quickly locating the fire is the most important step, so take advantage of knowing where the fire is located and move. With that said, quickly sweeping furniture enroute to the known fire location is appropriate and recommended. Being comfortable with splitting the search team will allow one member to be slightly more focused on searching for victims while the other searches for fire (searching for life and fire). For example, the firefighter carrying the water can goes straight down the hallway towards a bedroom fire. While the other firefighter, often being the officer, sweeps the couch, loveseat, and coffee table, working towards the fire room. Once the fire is located, then coordination between the engine company and primary search begins.

COORDINATE

The second step into gaining can confidence is to coordinate the engine company as they set up for extinguishment. The primary search team needs to be thought of and used as a reconnaissance team, giving an update on interior conditions and, most importantly, fire location. Radioing key information regarding the stretch of the first due engine company's hoseline may aid in extinguishment. An example of key information given over the radio from the officer of the primary search to the first due engine company officer, "Truck 61 to Engine 61, fire is located in a charlie side addition, extend 50 feet" or "Truck 61 to Engine 62, fire is in the bravo bedroom". Remember, firefighting is local so be sure to sub in your own department's jargon to ensure clear communication.

Although the search team has the primary responsibility of searching for life; searching for fire must not be completely forgotten. Once the fire room is located, the can firefighter moves to the fire room and starts a sequence of search and isolation (discussed in the next section). The officer should get a look at the fire conditions, confirm its location inside the building, and make a quick radio transmission to the first due engine company, coordinating that company. Once the radio transmission is complete and conditions allow, the officer should initiate a search starting with the immediate areas exposed to fire.

So, we have coordinated the first due engine company, but how does coordinating ventilation play a role in all this? It most cases ventilation, especially horizontal ventilation, should be withheld until the engine company has water on the fire. The water can has limited water with a flow rate that lacks "punch". This means an overly vented fire building has the potential to outrun the effectiveness of the water can. The water can preforms most optimally with ventilation limited conditions. Water will have a greater effect on the fire environment with limited oxygen. 2 ½ gallons of water, applied

in the appropriate location, will produce a profound knock back when there is little oxygen. Think of this, the water can starts the extinguishment process while the hoseline finishes extinguishment. This sets a reminder that the water can is used to knock back or isolate the fire, not extinguish the fire.

ISOLATE

The final step into gaining can confidence is to isolate the fire, protecting the search. Searching ahead of the hoseline becomes more optimal once the primary search team has taken control of the fire room. They do this by two simple tactics— apply water (from the water can) and closing the fire room door. These two tactics slow the forward progression of fire, which in turn, buys time for search and suppression efforts.

Closing the fire room door will reduce the amount of fresh air intake, even if you find the door partially burned away. A closed door slows the fire's intensification by staving it of oxygen. A typical interior hollow core door, with a twenty-minute fire rating, will hold back a tremendous amount of fire. These doors get their fire rating by the air space created between the two outer skins. Now, add up to

2 ½ gallons of properly placed water, and we have taken back searchable space.

A properly filled water can will have an initial stream reach of approximately 50 feet. We can use this stream reach when making a push down a hallway with gasses igniting overhead. This is a great way to cool and contract gases back into the room of origin. Once we get to the threshold of the fire room, the stream must be directed to the base of the fuel burning, this is known as base hit. Directing the stream at the base cannot be overstated, this is the difference between a great knock back or just "pissing in the wind". Do not spray water into open flame with the water can! What little volume of water the can has will be evaporated prior to landing on the burning surfaces.

Once the fire is knocked back with a base hit, if possible, look below the exhausting smoke and steam to see if there is a victim inside the fire room. Once the quick visual is completed, reach in and pull the door closed. Any water left in the water can should be used to wet the hallway side of the fire room door. This will increase the time it takes fire to fail the door. The can firefighter should continue to hold control of the door while the officer searches the area and/or bedrooms closest to the fire room. Typically, this continues until one of two things happens; one, the engine company arrives at the fire room, or two, the water can expels all water. In either case, the can firefighter should reconnect with the officer and join in the search for life. Now, let us bring it all together into a real-life scenario for total comprehension.

IT'S ON!

Your company has primary search in a single-story single-family dwelling with heavy fire out two windows in the rear. Looking over your right shoulder while riding backwards, you see the hellish conditions through the windshield. The rig is positioned, airbrake engaged, and cab door swings open. Stepping outside the rig, you noticed two adults between you and the burning home in a panic.

As you grab your tools, they run towards you hysterically shouting "My daughter is in there! My daughter is in there!". After acknowledging their screams, you feel an elevated heartbeat pressing against your bunker coat as you jog to the entry point. The weight of the irons and the water can feel noticeably lighter compared to carrying them on a fire alarm one hour prior. "Doors open!" yells your Lieutenant, informing you force entry is not warranted. Sweat is now running down your face while you tighten up the final SCBA facepiece strap.

It goes lights out as you start a life, fire, layout at the threshold of the front door. Getting below the smoke layer shouting, "Is anybody here?". Looking and listening for life; you see and hear no one. Looking and listening for fire; you see a faint glow down a hallway. Looking for the home's layout; you recognize the bottom third of a couch and assume a living room entry point. "Hey LT, I got fire in front of me!", you communicate with your Lieutenant. "Check. Go straight to the fire room", the lieutenant competently replies. While crawling in, you hear fire crackling as gasses are releasing from the burning fuel. Smoke is rapidly being pulled over your head as the fire breaths. Approaching the fire room, you hear your Lieutenant communicating on the portable radio. "Truck 61 to Engine 61, we got your fire room; bedroom off the hallway!".

The water can is pulled in an upright position to better locate the locking hitch pin. As soon as the pin is pulled you depress the handle and aim the stream to the flames above your head. You then quickly redirect the stream to the base of the burning fuels as the fire seemingly gets pushed back into the bedroom. Your Lieutenant grabs your SCBA cylinder to get your attention and tells you "I have a bedroom across the hallway, I'm going in for search". By this time, the fire has been darkened down, allowing you to perform another life, fire, layout of the fire room; you see and feel nothing from the doorway. To further protect the search, you close the partially burned bedroom door and direct, what little water you have left, to the hallway side of the door.

Here comes Engine 61, charging down the hallway. You hear them pushing in just as the water can starts to spit more air than water. Bam! The nozzle firefighter bumps right into your shoulder, almost knocking you down. "It's in there! It's in there!", you repeat with aggression. The nozzle bail

SECTION 4: ACTION

is pulled open as the firefighter on the nozzle makes the turn into the fire room. What little visibly you had in the hallway is now gone.

"Victim! Victim! Victim, Truck 61 has a victim!" you can hear being said over the portable radio. It's your Lieutenant, in the bedroom adjacent to the fire room, he has a victim. Leaving your empty water can behind, you dash into your Lieutenant's bedroom and close the door behind you. "LT, I'm here!", you shout. Finding the bed where a lifeless 9-year-old girl is being placed onto the bedroom floor. "Here's my Halligan, take the window", your Lieutenant yells. The sound of the window shattering overtakes all other fireground noises.

Due to the room being isolated, the smoke lifts to mid window height quickly. Without direction, you roll out of the first-floor window onto the grass. Two paramedics and a few members from another company are sprinting towards you with stretcher in tow. You turn and face the window and out appears the little girl draped across the arms of your Lieutenant. You grab her with your left hand as you're ripping off your helmet and facepiece with your right. Checking for signs of life; "She is breathing!", you shout. Paramedics take the little girl and begin treating her in the back of the ambulance. The noise of the ambulance doors closing; a unique sound that you've heard thousands of times before. A cloud of dust follows the ambulance as it races away from the scene heading towards the downtown hospital.

During the on scene hot wash, it was noted how profound the water can was in reducing fire spread and protecting the search. It allowed the Lieutenant to confidently search knowing the fire, which was likely traveling towards the little girl's bedroom, had been knocked back to the room of origin.

A few weeks later you get a special firehouse visit from the little girl and her parents. She has made a full recovery and wanted to bring the firehouse some homemade chocolate chip cookies. The little girl and her family received a VIP tour of the firehouse as the firefighters got to know them more. Once they said their goodbyes, you treated yourself to the cookies left behind. Straight away, you thought to yourself, these are the best cookies you have ever had!

FINAL THOUGHTS

Knowing how to use the water can effectively will increase confidence in its ability to provide protection while searching. However, to build confidence you must train. Just like any tactic, tool or fire ground skill, you must drill on it consistently. Figure out what works for you and what doesn't. Use these training drills to not only work on locating, coordinating, and isolating but also how to carry the water can while searching. At the end of the day, civilians trapped by fire need search and suppression executed simultaneously. Add a 2 ½ water extinguisher to the search team, operating ahead of the hoseline, and you will find a solution so eloquent it's optimal.

ORIENTED VS SPLIT SEARCH

GRANT SCHWALBE

"A busy company from a big city that consistently works together and sees a lot of work may function like a varsity football team. They will be able to speak less and spread out more to reach areas quicker."

Searching from the front door is generally accepted as the gold standard for residential primary searches. There are scenarios when the Hasty (Engine based) Search and VES may be used but let's discuss the primary search options from the front door.

I first must state for the record that I am not a fan of the TIC lead search from the front door. This is the type of search where the officer remains at the door and directs where firefighters search while using the TIC. I believe that since the TIC is a line-of-sight tool, the officer must be out in front and lead from different positions throughout the house. I also believe the TIC has limitations both in how it functions and our understanding of the technology.

My base level of training on search was the "Oriented Search" from Chief Skip Coleman (Toledo, Ohio). He taught that this was an officer led search, with the officer determining the areas to search and then having the firefighters search those areas. Very early I realized that the "Stay on the Wall, Search Train" being taught at fire academies was garbage and this was a much better way to search. Much later in my career I got to speak and train with a lot of firefighters from across the county. They proclaimed that the split search is the

way to go and that oriented search was slow and ineffective. What I found was that understanding and definitions on these of these two searches vary, this leading to the debate. In some areas you'd think these are as different from each other as republicans and democrats. I have a different take on that.

Which is better, an oriented search or a split search? I think an easier way to differentiate between the two is to compare it to Junior Varsity and Varsity Football teams. Both have a consistent set of rules, and the objective is the same. What differs is the level and speed at which they play the game. In search the big difference is how closely the searchers stay together and how much direction needs to be given. Let's compare the two types of search.

RESPONSIBILITIES

When a crew is assigned search, or any task, the officer or senior man directs the crew how to complete the task. They then must report back to the IC whether the benchmark has been completed, unable to complete or if more resources are needed. The junior member is often the "worker bee" spending the majority of his or her effort on the task.

For search we need to look at the responsibilities of each position. The officer is responsible for getting the task done, direct the order of areas to be searched, assure thoroughness (all areas get searched), set the tempo, maintain accountability and find a way out when a victim is found or once the search is done. In contrast, the firefighter is held accountable to be sure each room searched was searched completely, report and remove any victims to his/her officer and to know where they are in the building in relation to the officer.

- **Officer**
  - Tempo
  - Thoroughness
  - Accountability of crew
  - Situational Awareness
  - Safe way out

- **Firefighter**
  - Searching
  - Situational Awareness
  - Know where officer is

THE ORIENTED SEARCH

Searching from the front door is the preference for search if there can only be one search team. If the hose-line is already in place, follow the line to the fire to assure it is in place. The reason for this is that the oriented search team typically enters from the front door which gives easy access to the whole house. In addition, the majority of victims are found between the front door and fire and then the bedrooms. A search from the front door puts the searchers in the best position to find victims.

When determining where to search we want to get to the areas where victims are in the most danger and then where they are most likely to be. A good rule of thumb is:

- *On the fire floor you must go to the fire and then search out from there.*

- *When not on the fire floor the search should start at the point of entry.*

Once the fire room is clear (by us or the attack team) we move to the bedrooms and clear those rooms. By doing our search in this manner we assure the line made it to the fire, we cleared the front door to the fire (36% of victims) and then cleared the bedrooms (45-55% of victims). Oriented search clearly defines the focus of the searchers into two areas:

- **Officer:**
Leads the search and drops member into room to be searched while he/she decides where to search and sets the tempo. The Officer is responsible to set priority on what needs to be searched first and next, and to assure all areas searched. He/she is responsible to maintain orientation, plan exit if victim found, coordinate with attack, coordinate with IC, watch out for crew safety and to maintain crew integrity.

- **Firefighter/Searcher**:
Irons (Leaves striking tool at door). Searches from front door while officer locates fire, follows direction of officer and gets dumped into rooms as crew moves through house. Searches with hands, maintains tool or leaves tool at doorway of specific room being searched.

THE OFFICER LEADS

I am a firm believer that the Officer should be lead for searches. Having the officer lead better allows them to control the tempo of the search. The other reason is that most often times the officer has more experience which equals being able to more quickly recognize situations and what actions must take place. By recognizing this firsthand, it allows them to get directions to the firefighter. Consider the firefighter leading and trying to relay the situation to the officer. The officer must receive the information, make a decision and communicate that back to the firefighter. This leads to delays and an opportunity for misinformation. Leading simply allows him/her to set the pace and by leading we reduce the amount of communication that is needed. After getting a look for Life, Fire and Layout the officer should move towards the fire. If the attack team is in place, then the Officer should make contact with the Attack Officer and confirm:

- Attack has enough hose
- Water is on the fire.
- The fire area has been searched.

If any of those areas need addressed, they should be before the search team moves on. If the fire area has been searched by the attack team, then the search team must be sure to come back and get a secondary once the rest of the building has been searched.

If the fire room needs to be searched the officer should search the fire room while the firefighter searches outside of the fire room. If nothing above needs addressed, then the search team should move towards the bedrooms to begin their search. Once the bedrooms are all clear the search team should systematically search back towards the front door.

If the attack team has not already entered, the search team can report where the fire is at to the attack team.

Once the fire area has been searched the goal is to get to the bedrooms as quickly as possible. If

both members of the search team move in a line, they don't search a large area. If possible, they should be offset to cover more ground as they move towards the bedrooms. Once the bedrooms have been identified the oriented man literally dumps the searcher into the room to be searched. Then he quickly TIC's behind the searcher to call out layout and any victims. As the searcher searches, searches the common areas nearby, watches out for their safety and changing fire conditions but most importantly, identifies the next area to be searched.

One misconception with Oriented Search is that the Oriented Man does not search. This is where it's not black and white. The role of the Oriented Man is to direct the search and maintain an overall big picture mindset. The searcher goes in each room and completes the search. If the conditions are mild, then the firefighters can be further apart (voice or visual) and can "split search" or leapfrog rooms. This would be a case where the Oriented Man DOES actively search. If the conditions are bad or the searcher is new, the oriented man may choose to stay closer to the firefighter. In an area that bedrooms are clustered together the oriented person can search the hall and the bathroom while waiting for the searcher to finish the bedrooms.

Because the Officer/Oriented Man is maintaining orientation, this allows the searcher to concentrate solely on the search. He/she can search the room as fast as possible, knowing that someone is in the hall, within voice contact looking out for them. In addition, the searcher can leave the halligan at the door to the bedroom he/she is in. This will allow the searcher to use both hands to better identify objects in the room. If the tool is needed for whatever reason the Officer/Oriented man is close by and can bring it up.

The searcher must check behind doors, under beds, above beds (to check for presence of bunk beds) and in closets. It's usually not necessary to search walls unless you are specifically looking for a window to open or to use for egress.

STAYING TOGETHER

OSHA says we need to stay in contact one of three ways (since they are in a hazardous environment).

- Voice
- Visual
- Touch

Two things will dictate which method the searchers choose to stay in contact are:

FIRE CONDITIONS:
The worse the conditions, the closer the searchers will want to stay together. The better the conditions are then the further they can safely be from each other. If the goal is to search quickly this is best done if searchers can spread out.

TRAINING LEVEL:
If two seasoned firefighters are working together and have for a while, then they are likely to know what each other is going to do and search further apart from each other. If an officer has a new firefighter, then he will likely stay close to him/her so that the officer can keep a close eye on the new firefighter.

SECTION 4: ACTION

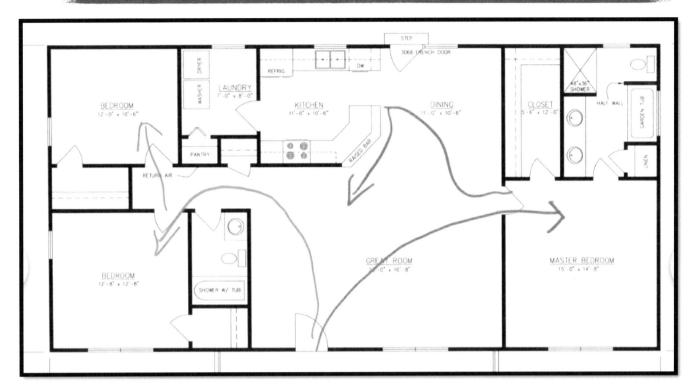

SPLIT SEARCH

The goal of the search is to quickly locate and remove victims. We can do this quicker the further apart we are. As listed above this comes with experience level of members and the conditions of the structure. As we spread apart, our search moves from an oriented search to a split search (if we insist on calling it something). The officer or senior member is still responsible for all what is listed above however in split search they may be able to do those things and be able to search as well. If conditions and experience allow for this, it will allow for the search to be completed faster-which is the goal.

SPLITTING THE SPACE

Medical calls are good opportunities to practice splitting the space. Once the patient has been taken care of, step back and have a quick talk on how you'd split the space. Some options for splitting depending on the layout are:

- Master bedroom-other bedrooms
- Left and right house split
- Left and right hall split
- Leap frog rooms

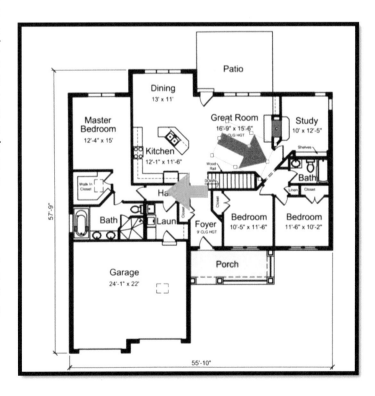

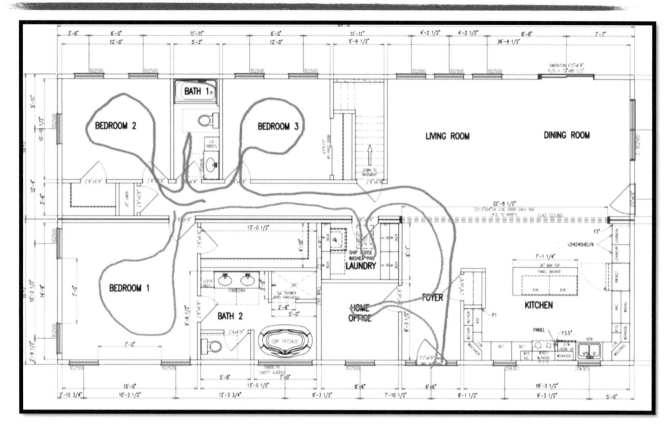

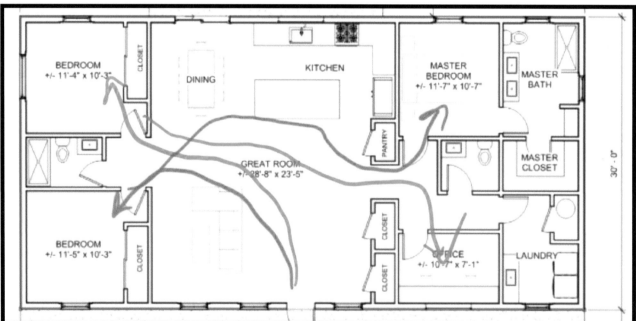

Getting a chance to talk about the split often will allow for less guessing on the fire ground. When split searching is more common it's not odd for both members to carry a bar.

CONCLUSION

When we debate which is better-an oriented search or a split search we miss the bigger picture. Both have their place and both are similar in how they are completed. The main difference is how closely the searchers stay together and how much direction is needed to be given.

The worse the conditions the tighter the crew stays. The better the conditions the further away they can get to speed up the search. Likewise, experience and training counts. The more familiar the crew is with each other and the more trained and seasoned they are, the further they can spread out from each other. This speeds up the search. The ultimate goal of the search is to quickly locate and remove victims. We can do this quicker the further apart we are. Train on our roles, understand when we can and should spread out and get those searches done quickly!

ORIENTED SEARCH
DAVE MELLEN

In the high-stakes world of firefighting, every second counts. Finding and removing victims hinges on performing a rapid, efficient, and effective search. Over the years, firefighting techniques have evolved significantly, and one crucial aspect that has undergone notable transformation is the oriented search. This search style is frequently mistaken for the split and often misunderstood as "just looking through the TIC". In order to better understand the oriented search, we need to examine the history behind it and how it came to be what it is today.

HISTORY LESSON

The concept of oriented search in firefighting traces its roots back to the mid-20th century. Historically, firefighters utilized rudimentary methods for search, often relying on instinct and brute force to search for victims. However, as buildings grew taller, wider, and more complex, traditional search methods proved inadequate for navigating through dense smoke and more difficult spaces. In the 1960s, firefighters began experimenting with new strategies to enhance search efficiency. One such tactic that quickly became mainstream was the Right hand, Left hand (RH/LH) search. Instead of blindly moving through smoke-filled rooms, firefighters learned to utilize reference points and maintain orientation within the structure by communicating with each other.

We have to remember that this was done in an era when radios were sparse and the idea behind a RH/LH search included the fact that if firefighters went into a burning building using a right-handed search and didn't come back out, additional search teams could follow the same path to find them. In modern firefighting where radios and thermal imagers are commonplace, this seems unnecessary, yet we still find organizations that consider RH/LH searches standard practice regardless of the area being searched.

Early search techniques primarily revolved around maintaining physical contact with walls or other structural features while navigating through hazardous environments. This was done in an era when radios were a luxury and a majority of the tools we consider commonplace today were non-existent. Firefighters would follow a systematic approach, feeling their way along surfaces to maintain spatial awareness and prevent disorientation. However, these early techniques posed several challenges. Limited visibility due to smoke and darkness often impeded progress, while the chaotic nature of fire dynamics made it difficult to predict where victims may be found.

Moreover, the reliance on only tactile feedback for orientation made it challenging to navigate through unfamiliar or cluttered spaces. This lead to the development of the oriented search. A systematic approach that allowed one firefighter to serve as the "anchor" and the other firefighter or firefighters to search an area relying on the direction of the anchor to help them with orientation. This allowed for much quicker searches and gave firefighters the ability to quickly transition from one area to another.

THE IMAPCT OF TECHNOLOGY

As technology advanced, so too did the capabilities of the oriented search technique. The introduction of thermal imaging cameras (TICs) revolutionized firefighting by providing real-time visual feedback in smoke-filled environments. TICs enable firefighters to visualize heat signatures and locate potential victims with greater accuracy, reducing search times and improving overall efficiency. This also allowed the anchor firefighter to quickly perform a sweep of an area and then guide the search firefighter to a specific area of interest. Take for example a room with a bed, an end table, and a dresser. The anchor firefighter could guide the search firefighter to the bed to perform a rapid but thorough search before moving on to a more systematic search for the rest of the room. In addition to technological advancements, training methodologies evolved to emphasize the importance of situational awareness and spatial orientation. Firefighters began using rigorous training exercises to simulate real-world scenarios, honing their skills in navigation, communication, and teamwork.

THE MODERN FIREGROUND

In more modern firefighting operations, oriented search techniques are integral to search and removal standards. Firefighters are trained to conduct a primary search upon entry into a structure, using tactile and visual cues to maintain orientation throughout the operation but also utilize the anchor to help facilitate a rapid search in areas of most probability to find a victim. According to the Firefighter Rescue Survey's First 3000 data, 16% of all rescues were made by using the oriented search. Furthermore, 87% of all rescues were made by finding the victim through eyesight, TIC, or touch. This number proves that performing a rapid, efficient search is key to finding and removing victims.

SYSTEMATIC APPROACH:

Oriented search technique emphasizes a systematic approach to navigation and focuses on maintaining spatial orientation relative to the structure's layout. This methodical approach minimizes the risk of becoming disoriented in challenging environments and allows firefighters to frequently judge where they are in relation to the point in which they entered.

TACTILE, AUDITORY, AND VISUAL FEEDBACK:

Unlike RH/LH searches, which rely primarily on sweeping motions to cover designated areas, oriented search incorporates tactile, auditory, and visual feedback to enhance situational awareness. Firefighters use their senses to detect changes in their environment, listen to directions from the anchor firefighter, and gain visual cues provided by TICs.

SECTION 4: ACTION

> **ADAPTABILITY AND FLEXIBILITY**
>
> Oriented search techniques are highly adaptable to dynamic fire-fighting scenarios. Firefighters can learn to adjust their approach based on evolving conditions, such as changes in fire behavior or structural layout. The oriented search can be done with two firefighters just as well as it can be with 4. This flexibility allows for rapid decision-making and effective problem-solving during search operations and gives firefighters the ability to rapidly transition from one area to another once an area has been searched.

Now that you understand the history and basic philosophy, let's examine the actual process of properly conducting an oriented search. The first and most important piece is understanding the role each firefighter serves during the oriented search. This is not absolute, and the roles can change from room to room, or floor to floor depending on conditions and needs. In oriented search the roles will be one of two positions, the "anchor" and the "search firefighter" or firefighters.

THE ANCHOR

The anchor position is most commonly the Company Officer or Senior firefighter. This is not a requirement however, by having the Company Officer or Senior member of the crew serve as the anchor, it allows them to maintain situational awareness of their personnel while simultaneously assisting them with direction. The idea of the anchor is to first and foremost, serve as a focal point that the searcher can reference to orient themselves within a compartment, usually at the point of entry. As the searcher moves through an area, they can communicate back and forth with the anchor and adjust their position based on areas of highest concern.

The next and equally critical function is to give the search firefighter direction and monitor conditions using a TIC. This will aid in expediting the search and prevent the crew from wasting time and energy searching an area that doesn't hold any value during a search. If no TIC is available, the anchor can still monitor the actions of the searcher(s) while listening and communicating with them by voice.

Several years ago, while searching the second floor of a home, one of my firefighters opened a door to a room that was completely empty. I searched all four walls with the TIC while they searched behind the door. With nothing found, we moved on. Imagine the amount of time and energy we would have used searching that room when someone could have been lying on the floor in the next bedroom.

THE SEARCHER(S)

The firefighters conducting the search are the workhorses of the oriented method. Under the direction of the anchor, personnel conducting the search can move into an area and conduct a rapid, efficient search while still maintaining situational awareness. The idea is to conduct a search methodically while focusing on areas that yield high potential for finding a victim. Take for example the searcher entering a room and the anchor sees a bunk bed, the area of highest concern at that point would be the top bunk. By sending the searcher to that area first, it allows them to search high and then low. After that, the searcher can resume a more traditional search using the anchor to keep themselves oriented.

THE PROCESS

Next, let's discuss the process in which an oriented search is carried out. The idea behind the oriented search is to break each area to be searched into individual searches. For example, if you are conducting a search on a hallway with 3 rooms, each room would be considered its own search. When starting at the entry into a new search area, it helps to identify the layout and use common terminology to identify areas of the room.

The first step is allowing the anchor to quickly scan the room with a TIC. This gives them a chance to get an idea of the layout, check for easily visible victims, find areas of interest, and check conditions within the room before sending the searcher in. When using a TIC, ensure that you are scanning all four walls, the floor, ceiling, and behind the entry door. This is just as applicable in a residential home

Photo by: Camera Worthy Photography

as it is in an office building or apartments. One word of caution is that the oriented search is not meant for searching large areas such as open layout commercial buildings, this type of occupancy would require large area search techniques. The oriented search is most applicable to residential homes, apartment buildings, hotels, office areas with cubicles, etc.

The next component is using common terminology. I like to use the "1234 method" of orienting the searcher. This is done by considering the wall in which you enter to be "Wall 1", moving clockwise are walls 2, 3, and 4. This reduces confusion when communicating about where the anchor wants the searcher to move. If I tell someone "I have a bed in the 2-3 corner", they know I'm talking about the wall ahead and to the left of them when they are at the door. This is the same numbering system we use when performing a VES so the firefighters are familiar with the terminology. By denoting walls in a clockwise fashion, it replicates the ABCD or 1234 system most commonly used for the exterior of a structure.

Each subsequent room or area will be treated the same. If a firefighter is searching and needs help or finds an area that leads to another room, a "Jack and Jill" bathroom for example, they can simply call out to the anchor who will send additional personnel or enter themselves to the next point of entry (ie bathroom door). Special caution should be taken not to overextend the search area. A good rule of thumb is that if you are out of voice range, you are exceeding the effective range of the oriented search. This may require a change in tactic to the split or large area search techniques.

Once the search of that room is complete the crew can choose to keep their positions or, in the event that the searcher has been exerting themselves, may opt to switch out with another firefighter, often referred to as "leapfrogging". This gives the firefighter who just completed a search an opportunity to take a position with less exertion and spread out the consumption of air more evenly.

IN ACTION

Assuming you have 3 firefighters (A,B, and C) where firefighter A is the anchor and firefighter B is searching, when the search of the first room is concluded, firefighter C would search the second room while firefighter B assists the anchor and prepares to search the next area. This works extremely well when searching occupancies such as hotels or office buildings where you may be searching extensively with limited staffing. If a firefighter becomes physically taxed, the option also exists to have them switch with the anchor so long as they are capable of performing that position effectively.

If at any point a victim is found, there are several things that need to happen immediately. The first and most crucial task is to call out "Victim, Victim, Victim" over the radio. There are only three phrases that have that cadence. "Victim, Victim, Victim", "Mayday, Mayday, Mayday" and inevitably when your mic is keyed up "F$&#, F$&#, F$&#". This gives everyone working with you the ability to stop what they are doing and assess your situation. The next crucial task is to ascertain if you need more help. If the victim is a child, you will

SECTION 4: ACTION

likely be able to move them by yourself however, if the victim is a 300 lb. dude, you will need help.

The anchor's job is to find out what, if any help is needed and send it. If no one else is available or you are a two-person search then the anchor can go to the search firefighter and assist. The last and arguably most critical task is to decide how you will remove the victim. If you are in a bedroom and there are windows, it may be a better option to take the victim out that way versus dragging them all the way through the occupancy. This will all need to be communicated to the IC and resources will need to be coordinated regardless of which way you decide to remove the victim.

Lastly, if the crew that was performing the oriented search is removing the victim, make sure to communicate where you left off with your search so that another crew can continue where you left off.

Nothing could be worse for a victim than starting the primary search over from square one when the next room that would have been searched contains victims. Remember, we are trying to minimize pauses and maximize efficiency and effectiveness.

Oriented Search in practicality:

I am a firm believer that you should not take advice from an instructor teaching you something they have never done. Those of you reading this chapter most likely don't know me personally and are relying on me being vetted by the other authors and Corley Moore himself. I want to end this chapter by sharing one example of how I used the oriented search that resulted in one of the most impactful grabs in my career thus far.

THIS IS AN ACCOUNT OF A FIRE THAT I RESPONDED TO ON THE MORNING OF MAY 13TH, 2014

As a Lieutenant, my crew and I were dispatched on a residential fire with entrapment. Initial radio traffic indicated that there was a female occupant who was trapped inside the home. On arrival we observed a single-story bungalow with heavy smoke coming from the Alpha side with an orange glow on the Charlie side.

We were assigned to search and immediately headed for the Delta side where the bedroom windows were located. My initial plan was to perform two simultaneous VESs on the bedrooms and find the victim. We were redirected before we got to the Delta side and told by the IC to enter on the BC corner where the kitchen was located. We entered the kitchen and made our way towards the Alpha side, searching as we went. When we reached the living room we observed fire dripping down from the light fixtures onto living room furniture and fire rolling across the ceiling.

My partner Chris and I made the decision to pass the fire and make our way to a T hallway. I entered the bathroom and Chris stayed at the door. I searched the bathtub and found the toilet by putting my hand directly into the bowl. I yelled to Chris that I had just put my hand in a toilet, and he chuckled and told me he saw it on the TIC. As I exited the bathroom Chris turned toward the front bedroom, and we switched spots. Chris entered the bedroom which was filled with boxes, a bed, and multiple other items. I directed Chris to the bed that he searched without finding a victim. He then proceeded to move around the room from wall 2 to 3 and then 4. Chris found a closet and I asked him if he needed me in the room with him which he yelled back "No, stay there". Once the front bedroom was cleared, we made our way to the back bedroom.

Once again Chris and switched spots to prevent either of us from going through our air quicker than the other. I opened the bedroom door and had approximately 1-2 feet of visibility under the smoke. I quickly scanned the floor with my flashlight and saw two bright pink house slippers with legs in them coming out of a closet. I made my way to the door and found an elderly woman confused and incapacitated. I grabbed her legs and pulled her out of the closet while yelling "Victim, Victim, Victim". Chris asked if I needed help, and I said no. She was 100 lbs at best and I felt I could handle her on my own. At this point I began dragging the woman to the door and I could hear Chris further than I anticipated yelling for me to come to the sound of his voice and to hurry up.

As I exited the bedroom both the victim and I were met with intense heat. I vaguely remember the victim yelling "I'm burning" so I turned my body toward the left where I felt the most heat. I then realized that the living room we had passed was now nearly fully on fire and visibility was near zero. Chris was yelling and I could tell his voice was to my right so I continued to drag the victim in that direction. As I crossed through the doorway to the front bedroom, Chris slammed the door shut, jumped over me, and was yelling by the window he had found. As I moved towards the window he called out "10 feet, 5 feet, 2 feet" because he was watching me in the TIC. We broke the window out and were met by a group of firefighters who helped pull the victim out.

The woman was transported to the hospital and later that day I found out that her name was Ann Elizabeth Jarrett, and she was 104 years old. Chris and I made our way to the hospital and as we walked into her room, Ann said "Those are my angels". Now I've been called a lot of things in my life, but angel wasn't one of them. Ann continued on to explain that the last conscious thought she had while trying to find her way out was "God, please send me an angel to rescue me" followed by a sense of warmth and then a bright light. Ann was scared to go toward the light because in her mind, she thought that was her demise and she wasn't ready. As Chris and I sat at her bedside I didn't have the heart to tell her that the heat she felt was from me opening the door and the light that she thought was her path to the pearly gates was my survivor light shining in her eyes. All that Ann cared about was that she asked for someone to come save her and we did just that. It wasn't heroic, it wasn't extraordinary, it was our JOB.

Without conducting the oriented search in the manner we did, I can't tell you that I wouldn't have gone into the bathroom which we found out later didn't have a window. The reason the search and removal of Ann went so well is because we had the flexibility to change on the fly and both Chris and I knew our positions interchangeably. Ann didn't care what color rig we showed up on, what brand SCBA we had, what search pattern we used, what Ann Jarrett cared about is that when she couldn't get out of her home on her own, we found her. Ann lived to be 108 years old and died in 2017 while taking a nap. She was sharp as a tack until the day she passed and I'm glad we were able to give Ann and her family those 4 years together.

CONCLUSION

The oriented search is one of many search techniques that you can choose from and will prove invaluable in the right situations. From its humble beginnings in the mid-20th century to its modern implementation today, oriented search has undergone continuous refinement and evolution that will continue to help you prioritize victim rescues.

SPLIT SEARCH
SEAN DUFFY

Photo by: Jillian Stewart - iShootFire

The assignment of search has two objectives, locate fire, and locate life. It is important to remember that searching inside a structure with an unknown fire location is the most dangerous position to be on the fireground. When searching ahead of the hose line and we locate the fire, we must communicate to fire attack not only the fire's location but the best-known access to the fire location. As we move through the structure, we confine the fire to create the best opportunity for us to search the largest area possible. However, the choice to use a split search is not just for when we are searching ahead of handline.

As searchers we have the privilege of being the person who is given the responsibility of making decisions for the incapacitated and the at risk. Citizens have given us the trust to make life and death decisions on their behalf. It is our duty to make the choices for their family, and their property when they are unable. We shall never forget the moral obligations we have to uphold their trust and make the decisions they would ask for if they had a say.

MAXIMIZE THE KNOWNS

If we know that the time it takes to locate and remove our civilians is a major contributing factor to their survival, then we need to be performing search in a way that not only allows for firefighters to cover more space, but also allows them the opportunity to find multiple routes of egress. Most rooms can easily be searched by one person so it's paramount that we use our staffing to our advantage during search operations. More staffing equals more rooms searched simultaneously! Split searches are not about free lancing and being reckless, they are about being intelligent and using our knowledge and experiences to our advantage. We need to be getting away from the "Flood the structure" approach to search operations where we just throw people into the building and call it good and not everyone has the staffing to throw multiple search crews inside a structure, at least not initially. There is a lot more to search than just checking the box and there is also a big difference between actually searching for people and accidentally stumbling upon them in the process of "Search". Our search is also not complete, just because a victim is located and removed. It is not uncommon to find multiple victims in fires located in or around the same vicinity as another. It is our job to search and find everyone, so our methods chosen should be aimed at doing such and in a respectable time frame. It is possible that by splitting we are able to find more than one victim simultaneously.

A split search will typically be performed when conditions are favorable, your crew is both comfortable and competent or fire attack is making a hit on the fire. A benefit to utilizing this type of search is that it decreases the search time by half and allows for civilians to have a greater chance of survival. Compared to other types of search, the split search is the most successful type in both overall number of victims located, and victim survivability. When it comes to primary searches of any kind having a plan is essential but even more so when it comes to split searches, as searching without a plan is like teaching without an objective or outline, you WILL miss something!

COORDINATION IS ALWAYS KEY

In a split search crews will split to target the bedrooms where there is a high probability of finding victims. During this type of search firefighters remain oriented by their initial search size-up, which can start long before the fire even occurs. The firefighters will simultaneously search bedrooms independently while also isolating the rooms they are in and ventilating. If isolation of a

Photo by: Jillian Stewart - iShootFire

room is not possible, fire attack MUST have a good knock on the fire before we can begin venting for life.

Following bedrooms, other search priorities will be the egress pathways as we move through the structure. Remember, we are looking for viable victims in obvious places! As someone who is assigned to search, we want to strongly consider human behavior in our search patterns. These are not definite that we will find anyone, but as searchers, we need to think like civilians. People tend to head directly towards the main routes of egress during emergency situations. In a structure fire people who are not aware of the fire or are trying to self-evacuate can be overcome by smoke and heat, this can leave them slumped over in chairs, lying in beds or on the floor near windows, doors or in the main pathways to egress. Children tend to go to their favorite spots when they are scared. These could be their bedrooms or siblings and parents' rooms because its where they feel safe. Be sure to be through and look in unorthodox places like play tents or other areas of concealment like piles of clothes. Children may also hide under or in their own bed as well as closets. If they are not able to retreat to their room, they may hide in other familiar spots to them. These spots are typically places where they like to spend a lot of their time.

WE MUST OVERCOME THE IGNORANCE

Most of us were not taught anything about split searching in our initial training and many of us may still not be allowed to perform it due to policies and procedures or we may not even know what it is as it is often misunderstood by fear. We typically are indoctrinated with the "You must never leave your partner" mentality which results in the choo-choo train style of search where we hold onto each other and don't branch out while always remaining in contact with a wall. Not only is this not effective but it ultimately decreases our orientation and is often very time-consuming.

WHEN DO WE DO IT?

So, when do we split search? When it comes to deciding when to split search there are a few things you may want to consider. One major factor to think about is conditions as conditions are a large contributor to what will drive out tactics. The best thing we can have on our side in a search is visibility. If visibility is good enough, you can maintain crew accountability through sight and sound which minimizes communications and distractions. Good visibility would be an example of favorable conditions and would allow for us to be more aggressive. What about fire attack? Many places will not even allow searches to begin until there is a hand line in place. In such case, a split search may be your preferred method. It is no secret that an aggressive Engine Company will set the stage for all other companies. Water is protection for everyone operating inside the structure and for our potential victims. Having this protection significantly lowers risk which allows us to be more aggressive in our searches. How comfortable are you and your crew with this technique? Crews that are well trained and experienced will be able to accomplish more than those who are not. This trust allows them to operate at a higher level, including the ability to branch further away from one another than those who are less comfortable and do not possess the knowledge or skills that they do. Having a disciplined and educated crew will allow searches to be completed quicker and more systematically.

In a situation where any one of these examples are present, it may warrant a split search. However, these are only a few considerations. As each fire scene is dynamic and presents its own set of challenges and conditions, our thinking and actions should also be adaptable to what will allow us to be the most effective. Training and a true understanding of street proven tactics is key to creating and implementing a strong search culture within any organization. Your firefighters want it and your citizens deserve it!

LIFE, FIRE, LAYOUT (AND CALL OUT!)

GARY LANE

Photo by: Camera Worthy Photography

In the past several years, more attention has been paid to the small details that can sometimes be lost to time within the large scope of firefighting operations and the other emergencies we all respond to. For the fire departments that have a well-documented history coupled with a strong tradition of passing on lessons learned, this may seem a moot point. For many other departments, however, paying attention to all the little things that the fire service throws at us on a daily basis, can easily find important information falling through the cracks. One such example is what we will discuss in this article.

Photo by: Camera Worthy Photography

A HISTORY LESSON

It's unclear exactly where, when and how the phrase "Life, Fire, Layout" became cemented as the go to verbiage for, what in essence, is simply a portion of our continuous size up during a fire. After several conversations over the years with different firefighters around the country, and names like Olson, McWilliams, Rosenbaum, and Avery being mentioned, there was one name that kept coming up regarding the phrasing. Cody Trestrail was that name. After speaking with Cody, he was able to provide a little bit of hazy background to the origins. He mentioned how numerous versions of similar phrasing were in use at least 15 years ago, and most likely longer, around the country. Cody worked on both large East and West Coast fire departments during his career (FDNY Squad 1 and Portland, OR. Engine 31) as well as traveling coast to coast providing hands-on training with his Brothers In Battle LLC. training cadre. Even he wasn't sure why the "Life, Fire, Layout" phrasing has stuck with so many, but fast forward to 2024, and it has become one of the most commonly used versions from coast to coast.

WHY IT MATTERS

What is this "Life, Fire, Layout" thing and why does it matter? Well, at its core, it's simply a cue to remind us what we're looking for, or looking at, during our personal size up when we pull up to a fire. It can start from the rig while still in the seat looking out the window. It continues as we are walking up to the house or building (watch out for "magic" as well… That's what Anthony Braxton with Portland, OR. Truck 7 has referred to when it comes to power lines sparking and arcing!) and then once we make entry into the structure. Rinse and repeat throughout the building as we hit different floors or intersections of rooms and hallways.

Let's break it down a bit more…

OUR FOCUS

As any fire department should be able to tell their new hires, our mission and reason for existing is to save lives. Our job demands that we put ourselves in uncomfortable, chaotic and dangerous positions at times, all while trying to save civilians that may be trapped by fire. Their lives are worth the risks we take. It is our sworn duty to step into the fray and make every attempt to help them, even at great personal sacrifice. The "Life" portion of this size-up focuses on two things: Could anyone be inside (almost always a possibility) and where is the best place to start looking for them?

Those familiar with the work done by the people behind the Firefighter Rescue Survey (www.firefighterrescuesurvey.com) that have been tracking and publishing data regarding civilian rescues over the past several years, know that we have found the hard evidence that backs up all the anecdotal "story telling" of so many firefighters over the years. We can now say with conviction, that people are being found at all times of the day and night, in all rooms and pathways of the house, regardless of cars in the driveway or "credible" reports stating someone is or is not still inside, and regardless of the building being reported "vacant" or not.

WE MUST KNOW THE DATA

When assigned to search, we must consider going with the highest percentage target... The bedrooms. Bedrooms are the number one area people are found, no matter the time of day or the neighborhood we work in. Whether we access them via the main entrance or initiate a window-based search, commonly referred to as "vent, enter, search" or simply "VES", speed matters, but we need to take advantage of the natural pauses that occur along the way. "Take the time to save time" if you will. That's where "Life, Fire, Layout" comes into play. Here's one example... Using the front door as our entry point, we may or may not have to force entry. Either way, when the door opens, be ready to control it if necessary, but also be ready to get low and take a peek inside and hopefully under the smoke layer if possible. Any signs of life? What about sounds? A faint cough might be our only clue that a victim is being overcome by smoke inside, just out of our sight and reach. The data mentioned previously from the Firefighter Rescue Survey has shown that as much as 10% of victims are being found within 6 feet of the door. Get low, flat on your belly if you have to, and take that look for life!

ADAPT AND OVERCOME

Another example would be during VES. After clearing the window, what is the visibility like? As we go over the sill and to the floor, again stop for a second and take in your surroundings. A little bit of lift goes a long way! 6% of victims have been documented right under the window! If that victim is a small child, you may be able to grab them and hand them off if anyone else is behind you at the window, or in the case of a first-floor window, just get them out and to the ground. If it's a larger person, then we may best be served by getting the bedroom door shut in an attempt to isolate the room due to the longer time frame and difficulty in getting an adult moved. This may not always be possible! Sometimes fire has taken hold of the hallway and has burned the top portion of the door off! This would be a glaring example of the "Fire" piece of that "Life, Fire, Layout" size up. Well, the fire has found us! Now what? Bail back out the window? Or could we quickly force another door off its hinges and place it over the half burned out one to buy us more time to remove a victim or even just finish searching this specific room? Have we trained and practiced this before? Sometimes it's easier said than done!

ANTICIPATION IS KEY

The "Fire" portion is hard to paint into a corner with a quick definition. Probably the biggest point to embrace is the LOCATION of the fire. Where is it right now and where is it headed next? "Locate, Confine, Extinguish" has been the gold

Photo by: Camera Worthy Photography

standard mantra for decades for a reason. Whether it's while we're pulling up, walking around or crawling into the building, the fires' location (and the conditions it's creating!) should be of utmost importance to everyone on the fireground. If you're the one to find it first or notice changes happening for the worse, make sure to let people know! A simple radio transmission that relays "Fire in the basement" or "Fire in the walls on 3rd floor" could make all the difference in how far we commit to certain actions in certain locations. An example of this might be finding a well-advanced basement fire that's burning holes through the first floor. This may prompt the crew assigned to search, to bypass the first floor all together (for now) and go all in on throwing ladders to upper floors and entering via windows to gain entry and start searching for trapped occupants.

GET AHEAD OF THE GAME

Regardless of the examples we could conjure up, taking advantage of the little 5-10 second pauses that consistently happen throughout the initial fire attack and search phase is critical. Dropping low, looking for life and fire and we see (and hear) nothing, but what about the layout? Having line of sight on the "Layout" of the room or hallway, providing a few inches of precious visibility off the floor, could be a game changer for our next actions. Guessing what the layout might be as we get off the rig and head towards the building, versus actually seeing the layout, is a night and day difference. What about when we make the second-floor landing of a house? Take a second or two, lay on the top step and look for that "Life, Fire, Layout" down the hallway. Maybe we see an orange glow down on the right, but also see that three other doors are all open into the common hall. It might not be a bad idea to go straight for closing every door first on your way to that last room the fire is in, do a quick sweep for life, and get that door shut if possible (even if it's only half a door, shutting it will help). Knowing the layout will speed up our decision-making process and allow us to act more swiftly.

DON'T FORGET TO CALLOUT

I'd like to mention another portion to this "Life, Fire, Layout" thing, and that's the "Call Out". Calling out into the dark something along the lines of "Fire Department! Anybody in here? Call out!" as we pause and get low, as we make that second-floor landing, or even after we have VES'd and are checking the hallway before closing the door (you did sweep the hallway, right?), could be the audible cue to a trapped occupant that help is here! When we call out, we need to follow it with a moment of silence and listen for a response. Please don't forget this part! It's easy to get wrapped up in breaking doors and glass, and plowing through rooms, but those things are very noisy, and that noise will cover the cough or moan of someone gasping their last breath 10 feet away from you if you don't get quiet for a second and listen! Sometimes a cough is all we may hear, but we have to be listening for it!

> *Arriving first due to a reported kitchen fire, and we're assigned search. We split our crew (2 through the front door and 2 going to VES the second-floor bedrooms) with the engine company pulling up right behind us. Dispatch said they received multiple calls confirming four kids, ages 3-14, are trapped upstairs. Stepping off the rig, a quick glance at the house style looks to be a 2.5 story wood frame, front door on the porch to the right (that's where I'll look for the stairs first) and I don't see any wires dropping and popping currently (no "Braxton magic", up front at least).*
>
> *Looks like the fire is all out the back from the glow, with just thick smoke pouring out the first and second floor windows. Breaking the irons loose as we make the porch landing, we take ten seconds and finish masking up (with gloves still on!). The door isn't locked, and we open it, drop to the floor and yell*
>
> *"Fire Department! Call out!" and then get quiet as we catch our breath for the next few seconds.*
>
> *No response is heard. Seeing the stairs are right where we thought, we make a beeline for the second floor as the engine company is calling for water right behind us. Laying on the top step, another call out, another pause, another listen. We hear glass breaking in a room ahead of us (hopefully that's the other two guys VES'ing) but taking an extra second to look at the hallway floor I recognize a child's feet in her jammies!*
>
> *"VICTIM, VICTIM, VICTIM!"*

SECTION 4: ACTION

> *I shout as I grab her and hand her to my can position firefighter right behind me, and off he goes knowing that every minute she's not breathing is a 10% drop in her chance of survival (per the AHA).*
>
> *Making a move further down the hall I come to a closed door. I open it and again call out "Fire Department! Where are you!?" and silently listen for any response. It's almost a shock for a split-second hearing a cough a few feet off to my left.*
>
> *I move that way and immediately find the other children. I call out "VICTIM, VICTIM, VICTIM" again. Luckily the VES team heard it, and they yell back they're coming to me. The children all get taken down the stairs which has started to lighten up a little bit with the engine crew making a knock on the fire in the rear kitchen and hydraulically ventilating with the nozzle out the window....*
>
> *It's been two and a half minutes since we got on scene.*

BUILDING BLOCKS

Painting a picture like this is what we need to envision when discussing "the job" with our peers. I don't care if we make one fire this year or one fire every shift for the year, we need to have our head in the game and a plan to go with it. Learning as much as you can about the district you serve, the types of buildings and the people that inhabit them, has got to be a daily priority to stay on top of. Add that to a couple of seconds at the rig making a quick size up and plan, can go a long way in keeping us on track and minimizing wasted actions and energy. Taking advantage of the pause when you find yourself stopping to reorient or to simply catch your breath, is another chance that can be used to get low and take a peek under the smoke or even just getting quiet and listening to your surroundings for a second. All of these build upon the "Life, Fire, Layout" concept and increase our chances of a successful operation.

IN CONCLUSION

I hope some of this discussion and these descriptions have helped, and maybe even sparked an interest in making "Life, Fire, Layout, Call Out" a part of your personal size up when it comes to operating at your next fire. All of these things can add up to higher performance from our members as they work their way through the chaos trying to make a difference to those in danger. Regardless of what you call it, paying attention to the details and taking these actions can, has, and will save lives. In closing, I'd like to paraphrase the words of the late FDNY Lt. Andrew Fredericks, who we lost on September 11th, 2001: "Expect fire and expect victims on every run".

THE SUBURBAN ENGINE SEARCH
SHANNON STONE

Photo by: Jillian Stewart - iShootFire

ORDER OF ACTIONS

Unlike our urban counterparts, suburban-urban fire department's functions on the fire ground can be sequenced (arrival) based and tactical tasks are largely determined based on the immediate priorities upon arrival. This primarily happens because most suburban fire departments are not fortunate enough to be heavily staffed and run double-company firehouses. That being the case, there can be a delay with companies arriving simultaneously at a fire. This places a unique challenge on first-due engine companies when faced with a working fire and possible or unknown entrapment. Having a delay in the second due company presents a potential detrimental delay in initiating an aggressive primary search. Unless the first due company has reliable intel on the location of a victim where a rescue can be made without the protection of water application, an aggressive advancement of the hose line to begin suppression is usually the best option. History will tell us that blindly searching with no intel and without some form of protection from an engine company can be catastrophic. This leads us to the "Engine Search".

THE ENGINE SEARCH

The "**Engine Search**" can simply be defined as searching survivable spaces while advancing the initial attack line to the fire. The engine search is not designed to replace an assigned primary search but rather to supplement it. This is a very dynamic function that relies heavily on the fire ground experience of firefighters. Conditions must be favorable to conduct the engine search. There are certainly times when this cannot be done due to the fire conditions; however, I would argue that most of the time some form of the engine search can be accomplished. Thanks to the efforts of many with the Firefighter Rescue Survey, we now have data to reinforce this important task. Roughly 76% of victims rescued occur by the first due fire company and 29% of all rescues occur when no report of occupants are trapped. When there is a delay in initiating a primary search simultaneously with the fire attack, the engine search should be the default.

HOW IT WORKS

As the attack line begins to advance, when feasible, the company should focus on searching the searchable areas from the entry point to the fire and from the nozzle back to the entry point. This is accomplished very quickly as the line begins to advance. Many times, the entry point is the front door of the home. If the fire is located in a bedroom or distant from the entry point, it is not uncommon to have decent visibility (2-4-8 feet) upon entry. A quick search with your eyes and a good handlight can clear the first few rooms quickly (living room, kitchen as an example). Scan searchable areas such as bathrooms, bedrooms, and closets while advancing the hose line. This can continue as the line advances until conditions deteriorate. At this point, the primary focus should be line management and advancement to place water on the fire. When to make this transition is an experienced-based judgment call and is also staffing-dependent. A three-man company (nozzle, officer, doorman) is more effective in conducting an engine search than a two-man company (nozzle, officer). Placing water on the fire immediately increases survivability and creates survivable space. DO NOT forget to always check behind the door. Victims have been missed many times before because this area was not quickly searched.

Photo by: Jillian Stewart - iShootFire

IT ALWAYS DEPENDS

The big question that many struggle with is, when a victim is found, does the nozzleman stay on the nozzle or assist with victim removal? Well, the all-famous answer is that *it depends*. As a rule, the nozzleman should never come off the nozzle unless the officer is 100% confident that a legitimate knockdown on the fire has been accomplished. The nozzleman is the company's only lifeline for protection from the fire. Fire burns differently than just twenty years ago and one of the major changes has been how quickly fires will rebound compared to years past. Hydrocarbons and other synthetics are a serious concern and will facilitate a rapid-fire rebound, often catching firefighters off guard. The officer must conduct a quick interior size-up and determine the best course of action. This decision will be based on two very simple things: training and experience. I cannot overstate the importance of understanding modern fire behavior and gaining practical experience to help with time-compressed decision-making. If the officer decides to have the nozzleman assist with victim removal, a solid tip is to have the nozzleman flow water for an additional ten seconds (even if you don't think it is needed) to completely water map the entire area to ensure complete knockdown has occurred. I would rather float a couch than burn a firefighter. A couple of things to consider when removing a victim is to train realistically. If you talk to any firefighters who have removed victims from house fires, they will tell you it is much more difficult than one would think. Although training with manikins helps, it will create training scares. Rest assured, removing a human being is completely different from removing a manikin. Train with real people under realistic simulated scenarios. Understand simple drag techniques that one and two firefighters can use. Understanding these techniques will help keep your nozzleman on the nozzle for your protection. The original entry point is not always the best route to remove a victim. Remove them the quickest and easiest route out of the structure. Over-communicate with your partner to ensure everyone is working in unison. The lack of communication will quickly make a task very difficult.

CONCLUSION

In closing, my personal experiences over the last three decades working in suburban/urban interface using the engine search has been extremely effective with many citizens rescued by the first due engine company. Again, this should in no way replace the primary search but understanding that when a fire attack and search cannot be done simultaneously due to lack of staffing, the engine search should be the default. Everything we do on the fire ground supports the aggressive search. Without question, it does work and saves lives.

TOWER LADDERS: RESCUES & REMOVALS

MIKE CIAMPO

Photo by: Camera Matty Daly

Tower ladder buckets provide a stable work platform for firefighters to work out of at various heights. This enables them to perform rescues and removals of numerous people or Vent, Enter and Search operations. When arriving on scene, lining up the bucket with the victim will reduce the travel time it takes to reach them. That differs from an aerial ladder's procedures where we line up the turntable with the victim. Buckets come in all different sizes and configurations and depending on these items, it will determine your tactical gameplan.

POSITIONING

On apparatus with front outward-opening doors, most chauffeurs like to line up the turntable, so the bucket is square to the building. When doing so, the bucket operator must leave enough space between the bucket and building, so the door can swing open and isn't stopped by the structure. On some of these buckets, the doors swing open further than the front lip and caution must be used when entering or exiting the bucket. A gap will exist between the bucket and building, which can cause a firefighter or victim to slip and plunge into the building, bucket or to the ground. In some cases, the bucket can be raised to a level where the door can open into the window and permit a safety barrier while entering or exiting the bucket. This procedure must take into mind: the size of the bucket door, the window size, shape and ornamental trim work, along with fire, heat and smoke conditions.

Photo by: Camera Matty Daly

VICTIM SIZE-UP

One of the most important things the bucket operator must do is a size-up of the victim's mental status, physical condition and position in relationship to the fire or smoke's location. Once that has been accomplished, we can decide on the method we are going to take in approaching the victim. Approaching the victim from below isn't ideal, this gives the victim the opportunity to dangle, jump or drop a child into the bucket when they feel it is close enough. Depending on their size, they could land on you, causing an injury or damage to the rig's operating system from the impact load. It's more practical to come up to an equal level, far enough away from the victim and then rotate toward them to make the rescue. During this period of operation, the firefighter in the bucket must be in constant verbal communication with the victim, informing them of their intent and actions. Don't be surprised at any time that during one of the movements, the victim makes a lunge for the bucket. This method works well when approaching a victim in a window or sitting on a window ledge.

PROACTIVE & REACTIVE

There are a few times when we may have to adjust our approach, which may include operating around or through overhead obstructions such as wires and tree branches to reach the victim. At other times we may have to come in from below when a person is hanging onto the windowsill with their arms or straddling the sill and rolling outward. During our approach, we'll notice that they are probably only holding on with their hands, legs and brute strength and we must act quickly. This may be the only time we should come in from below a victim, in case they were to suddenly lose their grip and plunge downward. If possible, in this scenario, try to bring the unoccupied section of the bucket toward their direction so if they drop, they don't land on the bucket operator. In the case of a child being dangled out of a window by a parent, we'll have to size-up the situation to see if coming in adjacent to them versus from below is the proper tactic to follow.

COMMUNICATE INTENT!

Some firefighters advocate coming in higher than the victim and then lowering the bucket into position to make the rescue. Often, people will think you're bypassing them and will get agitated and change their positions (possibly standing up in the window frame) and contemplate jumping for the bucket. There's also been times a person on a sill or balcony has started climbing over the railing going for the bucket that's higher than their position. Verbal communication must exist between the victim and rescuer in this situation; keeping the victim in position until the bucket arrives at their location is our primary goal.

When we place the bucket to a fire escape or balcony, having the floors line up usually allows most people to easily transition over the railing and into the bucket. If the building's balcony juts off the

SECTION 4: ACTION

Photo by: Jillian Stewart - iShootFire

structure, place the bucket to one of the sides. This enables the victim to have the building wall in front of them, which allows you to pin them if they slip. Coming head on into a balcony doesn't offer this tactic if the victim slips. In addition, with front bucket controls the box makes removals more difficult. Removing wood or lightweight metal balusters can allow the victim to sit down and slide downward into the bucket. Size-up the balusters and bring a portable saw or cutter if you encounter this situation.

MULTIPLE VICTIMS

There may be times when we face multiple people who need to be removed from a structure. Delivering all the victims down to the ground may not be our best option. When time is crucial and victims are exposed to smoke and fire, we may have to come up with other plans. Removing these victims to a lower floor, opposite wing of the building, another balcony or fire escape, the roof of an adjoining structure or any other safe area of refuge might be our best option. It may not be possible, but if we can have a firefighter remain with the group to reduce their panic levels it may be a wise thing to do. If that is not possible, notify the incident commander and have them direct personnel to accomplish that task.

If multiple victims need to be removed, one is being administered medical care or the firefighter in the bucket is carrying a young child, the pedestal operator should perform the lowering operation. With some tower ladders, the controlling levers on the pedestal offer more smooth and precise movements of the boom, making it a more fluid removal. This procedure can also be used when a victim is being cared for in a Stokes basket removal.

VENT - ENTER - SEARCH

Vent, Enter and Search operations can be completed into numerous areas when a structure is on fire from the safety of the tower ladder's bucket. Firefighters should always strive to operate with two members, but it is understood that at some times this is unachievable. A good tactic to follow is for one member to remain in the bucket to be: a safety guide, able to accept a victim or enter and assist in removing the victim and able to reposition the bucket, if necessary, while the other member performs the search of the room. If your truck company carries an additional thermal imaging camera, the firefighter in the bucket can monitor the conditions and search efforts. When staffing conditions warrant and only one firefighter is available to perform VES, they shall inform their officer or the incident commander of their intentions over the radio. A firefighter should be assigned to back-up this member as soon as possible or the Rapid Intervention Team should be aware of the location and continually monitor the search's progress. Also, dependent on the structure's type and occupancy or multiple victims reported to be in one room/area, two firefighters can enter the room and follow opposite walls to complete the search quicker.

ACCESS CONSIDERATIONS

Placing the bucket to a window for access will often depend on the bucket's approach to the window and where the bucket doors are located. Normally, the bucket should be positioned to the window, so the firefighter can transition onto the sill in an almost sitting position. Most times, the bucket's floor is a bit lower than the sill. Positioning it here allows the firefighter to be low in the window, out of escaping heat, smoke and gases. They can sweep and sound the floor with their leg to ensure floor stability prior to entering. Each individual firefighter's body style is different and finding the right height for entry is up to the member entering to search. Entering the room in this manner is a more controlled state, but there will be times when firefighters will choose another method of entry. After sweeping and sounding the floor for victims and floor stability, the firefighter may choose to enter with a head-first and slide down over the windowsill maneuver. All attempts should be made to control your speed when doing so, a large impact load could cause lightweight construction to be compromised. In addition, we don't want to just "dive" in and hit an object and go flying, injure a wrist or land on our tool. If the bucket's top rail is placed at or just below the sill and the firefighter is going to boost themselves up and over the railing, use caution; flashlights, bunker gear accessories, SCBA straps can get caught between the sill, the bucket railing and building when entering in this fashion. In addition, ensure the power off button is engaged on the bucket control box, so while entering or exiting it isn't inadvertently struck, and the bucket moves.

Photo by: Camera Matty Daly

KEEP IT COORDINATED & STAY ORIENTED

When performing VES operations, notify units of your intentions of venting so they can be prepared for the introduction of fresh air. After venting the window, let it "blow" for a few seconds, watch and read the smoke conditions coming from the window. Is it violent, under pressure and swirling (ready to ignite) or is it lazy and lingering smoke (door is closed to the room, are you further from the fire)? Do you feel moderate or high heat exiting the window? Be cognizant of the conditions you're about to enter and continuously monitor them once you're inside. Prior to entering the room, if the bucket is equipped with a portable light on an electrical cord, it may be placed into the window for a reference point. Some firefighters will also slide their 6' hook into the room to assist them entering and to find the window on their way out. The hook can also be used to scan the walls about midway up, looking for other windows or trying to close the room's door to isolate the area from the approaching fire conditions from when you first enter the window. This tactic will buy firefighters time when isolating themselves from the fire's extension and decrease drawing the fire into the room being searched.

VICTIM REMOVAL

When a victim is found and must be removed, there are a few ways to complete the task and there is no set rule, due to the bucket's position to the building. Yes, we can say we favor a specific way, but when we encounter obstacles or obstructions, can we get the perfect bucket position all the time? That answer is no, so we should learn a few methods of doing this and train on them. If we are operating all alone and the victim can be lifted out the window and into the bucket, we'll most likely place the victim into the bucket through the door we entered from. Lifting the victim up by the torso and setting them up on the sill is our best option for controlling and removing them into the bucket. Sometimes if the victim is larger, we can set them halfway up on the sill and climb or crawl over them into the bucket. Now we can either drag them into the bucket through the doorway or lower the bucket a little to let gravity help us remove them. Just use caution when they're coming over the sill, their speed will pick up and a rapid movement may occur. The one thing about lowering the bucket slightly, it may allow the firefighter in the bucket to lift the victim up a little and walk backwards in the bucket as the victim comes across the sill.

Of course, if we have another firefighter in the bucket, they can adjust bucket positioning for the best removal height possible. They can decide which option is best for removal by sizing-up the victim's size and body position as they are coming out the window. Another option is to drop the bucket with the top railing below the sill, so the victim glides easily out the window and down into the bucket. Use caution again, the speed of the victim coming down into the bucket can be rapid. Plus, the victim can already have medical issues, and we don't want to cause any more. Always use caution if you're accepting a victim over the railing, each bucket is different and there could always be a chance their body slides on the railing. If we're not in a perfect operating position, which can happen, we don't need a victim sliding off the side of the bucket. Also, if the victim is coming over the railing, ensure the bucket doors are closed; one, it provides an enclosed area and it won't cause a firefighter to fall out of the bucket. If you can only remove the victim over the front railing, be advised that bucket control devices should be in the off position or slid to one side of the bucket. Plus, the master stream devices may keep you further away from the structure, creating a gap, so use caution during the removal process. If the bucket is below the sill with an awaiting firefighter, a smaller victim handed out will be in a cleaner atmosphere once they are removed from the structure.

CALL THE AUDIBLE

If a member is performing VES from the bucket and an imminent life-saving measure must be performed in another area, the bucket can't be moved until the searching member is notified over the portable radio. If he must abandon his search, he should return to the bucket and then perform the other rescue. If they're not in any imminent danger, the bucket can be moved to make the rescue and then as soon as possible return to the window where the member was performing VES operations.

DEPLOYMENT OF VICTIMS

The victim is not considered removed until they are on the ground. With many midship tower ladders we'll be able to place the bucket onto the ground at either side of the apparatus and facilitate an easy transition from the bucket to the ground and awaiting EMS units. There will be times when this is not achievable, and the victim must be removed from the bucket or the turntable and down the apparatus access stairs to the ground. If they can assist in their own removal, a firefighter should still accompany them out of the bucket and to the ground due to their unfamiliarity with the rig's stairs/ladders or due to their physical condition. In the event of removing young children, they may be handed down from the bucket, upper deck or the turntable of the apparatus.

If victims are in the bucket of a rear-mount tower ladder and the boom has been re-cradled due to no clearance on the sides for the bucket to be placed to the ground, we'll have to revert to another way to remove the victims. When victims are agile and mobile, they may be able to walk down the fly sections of the ladder toward the turntable to complete removal to the ground. Another option would be for a firefighter on the ground to raise a portable ladder (Folding/Suitcase, Fresno, or combination A-frame) up to the bucket and have the victims climb down while being supported during their descent.

SEARCHING OFF THE INITIAL ATTACK LINE

BOBBY ECKERT

Photo by: Jillian Stewart - iShootFire

The Initial Actions of first arriving units at a working house fire have a deeper impact on the incident than anyone arriving after. The actions of those Firefighters can not only stabilize the incident in question but perform single tasks that can contribute to life safety. Simultaneous actions need to unfold no matter what. There's no doubt that water to the seat of the fire is going to be the BEST thing for not only our people but any who may be lying unconscious in the building. One of our major problems is we DON'T know where these people are and we need to get to them, as quickly, if not sooner than the water arrives to the seat of the fire.

SKILL MATTERS

As with anything in this trade, it will always fall back to training. The more proficient we are with getting that first line deployed, than everyone as a fighting chance. We always need to ensure we select, pull, and place the proper hose line. If we're going interior than speed and mobility need to be in the forefront of that decision. Water to the seat as soon as possible. Being on an Engine Company or assigned to the first hose line does not mean we are passing the search off to another crew. That search begins as the push is being made. Anyone in that building overcome could be encountered on the drive to where that hose line needs to be placed. A single Firefighter needs to be trained to not only deploy but get to the objective (fire) by themselves with zero help. Our focus in Firefighting should not be what we don't have— BUT what we do and making it work for us. The great thing about house fires is the compartmentation not only aids in us getting into position— BUT can aid a person overcome lying in survivable space!!!

GIVE THEM A CHANCE

As a Company Officer I've found great success by finding the location of the fire and guiding the nozzle firefighter to where they need to be and searching along the way. Once they're in position, if the fire is giving off any visibility (which at times it will!!) to tell them to hold off before opening that bail. Once that nozzle is opened any visibility that existed is gone until fans are placed in the building after fire attack is complete. Use any of that visibility to your advantage to scan the immediate area of the fire room and anything adjacent to it. After that nozzle firefighter is set and has water flowing, and there's no kinks, it's time to peel off. Start at their position and which is directly, at the seat of the fire and start working your way back. Move towards the compartmented rooms immediately adjacent to the fire room (s). While searching keep in contact either voice or radio with the nozzle firefighter. Never moving too far from the nozzle making the push. This move can take minutes and may have dividends for anyone who is overcome. It's important not to get too far from the nozzle firefighter, but yet give yourself a chance to hit those immediate areas that may pose the most danger to any civilian.

Photo by: Jillian Stewart - iShootFire

OUR MISSION

Being on the engine company does not make any interior firefighter in that Company immune to searches. Anyone performing Truck Company functions may be overwhelmed with conditions upon arrival and may have to perform tasks that delay searches. There are many Engine Companies who handle their own entry due to the truck being delayed or busy with other functions. No matter the operational reality we operate, we need to make the proper adjustment so simultaneous tasks are being completed along with the main fire suppression. As with anything this evolves with training on what's at our disposal and making it work for us in an array of situations. As with anything this tactic may not be optimal, for example a basement fire, where it's more challenging getting that first line in place. BUT when that line is in place and flowing, get that search started. The unpredictable, along with zero visibility is our work environment. BUT we have training, SCBA, turnout gear, tools, and water. Their best chance is us. Even if that is unconventional to what standard fire service texts say. Experience driven tactics matter, because they not only work but have positive results.

Stay Effective.

SECONDARY SEARCH
JON QUON

Photo by: Jillian Stewart - iShootFire

Imagine primary search and secondary search in the movie Twins— secondary would-be Danny DeVito, to the primary's Arnold Schwarzenegger! In fact, it doesn't seem to be given much attention by anyone, an afterthought to an afterthought. When it comes to information on secondary search you are hard pressed to find more than a paragraph in most texts. These meager offerings don't offer us more than the usual "thorough search performed by a separate crew after the fire is controlled" platitudes that we have come to know and accept. To be clear this author's relationship with secondary search has been no different. My engagement with secondary search, or lack thereof, is made all the worse considering how much time, effort, and passion I bring to the primary. It seems only right that before we discuss the secondary and all it entails, we define its purpose. Secondary search exists to ensure that all spaces of the structure that were searched by the primary crew and any that were missed or unsearched due to the hostility of the conditions, are in fact, negative for victims.

FOLLOW THE DATA

The Firefighter Rescue Survey (FRS) has shown us that a victim's best chance for survival lies in a total removal time of less than 2 minutes (79%) and that it drops precipitously when total rescue time takes longer than 6 minutes (65% survival for 4-6 min total removal time). The outcomes are even more grim when it takes 15 minutes or greater to find and remove the victim (47%). Of the initial 3,000

Photo by: Jillian Stewart - iShootFire

recorded surveys, a victim was discovered by the secondary crew only 4% of the time. That number may seem inconsequential, and it does pale in comparison to the 57% discovered by the primary, but that in and of itself is not a bad thing. It means our primary searches are finding the victims as they should, but it also reminds us that by their very nature they must be executed with such urgency that a victim can be missed. To perform the search at a lesser tempo would in fact not be in the potential victim's best interest given the truncated time window with which we have to seek a positive outcome. The 4% who were discovered by the secondary search represent to me 120 souls who were given, if not a chance at survival, a small measure of dignity and closure to their families.

During the secondary, the tempo should be reflective of the possibility of locating life. Outliers exist when it comes to survival, and it is not our place to make judgement on where life can persist. Bearing this in mind, it is now our duty to ensure there is no chance a victim can remain in the fire building. In addition to all living spaces in the structure we should be looking to lay eyes on knee wall voids, closets that may have been obstructed by debris or excessive contents and missed. Many of the older homes in my city have servants' stairways or stairs that access a half story through a closet. Being familiar with the building stock in your area and common layouts will increase your index of suspicion for these unique features.

UNDERSTAND THE WHY

We have all heard that a secondary search should be performed by a crew other than the one who made the primary. But why? To paraphrase the great Tom Brennan, a room searched and found to be empty will remain that way in the mind of the searcher. Given a unique set of circumstances or limited staffing the luxury of a completely different crew may not exist. Consider this scenario: A four-person crew splits and 2 take the upstairs while 2 take the downstairs. Reports are given of victims on floor 2, if the initial search comes up empty and the potential victim is still unaccounted for, then it may be prudent to have the crew flip-flop. While not a true secondary, given the report of a victim, a fresh set of eyes is in the potential victim's best interest. I recognize much of the country will does not have four-person staffing, however we must consider these types of scenarios. Another scenario that could present itself is this: a primary has been completed and a homeowner arrives on scene stating a victim is in the house, is a new crew assigned search? Or does the same crew return and search again? We often skip the importance of a face to face with the primary crew whenever possible. During this exchange the primary crew can share the conditions their search was performed under, areas that were inaccessible due to fire, unique building features that hampered the search, level of contents in the structure, and any other pertinent information. This brief conversation may also offer a window into the quality of the primary search performed. The fireground can offer many forms of friction that affect the primary, as a secondary crew we trust but also verify.

BE DELIBERATE

During the secondary every pile of debris must be investigated, drywall that has accumulated on the floor during the firefight can obscure a victim. To illustrate this point, not long ago a deceased victim was discovered during the secondary in the city where I work. The body had been covered with drywall that had fallen/been pulled during the firefight. This combined with the large amount of contents in the room made the victim difficult to locate. Piles of clothing, and other contents must be sifted through. The same techniques we utilize during our primary i.e., not patting with our hands, but instead sifting for weight and contour should be utilized. If it was not possible during the primary to vent for search now that the fire is under control, this should be done to increase visibility. The outside of the structure should also be searched for jumpers that may have been hidden by overgrown bushes or obscured by the contents of the yard. On the commercial side of things all elevators and shafts should be searched. So, when does the secondary happen? I can hear you now: after the primary of course! This is true. Of equal importance, the fire is now under control, meaning that the conditions in the structure have improved. Recall, our primary aim is discovering every victim. Nevertheless, the devastating effect of fire can drastically change their appearance, often in a gruesome manner.

LEARN FROM EVERYTHING

Allow me to share a brief story that has impacted the way I view the importance of the secondary. Not long ago a crew from my job arrived at a house fire. They were met by frantic parents in the yard who gave them a location where a young child was trapped. This crew did everything within their power to find the child under severe and punishing conditions. They could not locate the child in the location given to them. During the secondary the child was found in the room of origin. Speaking with the officer, he shared with me that knowing that not only did they not miss the victim, but in fact it would have been impossible for them to reach them, was a small comfort in this difficult and challenging situation. As a secondary we may be able to absolve our partners in the primary of this moral injury. The fireground is a harsh and unforgiving place and occasionally despite our best efforts we will fail. Examining these failures, learning from them and sharing lessons learned is the way forward.

TAKE IT ALL SERIOUSLY

Consider this, the mark of a true professional is the way they embrace any task big or small. I myself have been guilty of coveting the higher profile fireground tasks. Let us all strive to bring the same level of intensity, care, and dedication to the tasks that have been previously considered less desirable on the fireground. Doing so will keep us on the path to becoming the quiet professional we all aspire to. The secondary is more than a box to be checked, it is a vital part of our mission which is to protect life. Maintaining a strong connection to our "why" will strengthen our resolve and bind us to the highest standard.

THE AUTHORS

A lot of amazing people contributed their efforts to making this work a reality. In the following pages you can find all the authors who wrote an article.

They are listed alphabetically so you can easily find an author and what article they contributed.

AUTHOR GLOSSARY

IAN BENNETT
TRAINING FOR SEARCH – Pg 55

Ian Bennett is a Captain with the Seattle Fire Department. He has been involved in training since 2010, including teaching multiple recruit classes and developing and delivering multiple classes for operations members. He was part of the team that formalized the SFD's live fire training program. He is a former instructor at the Washington State Fire Training Academy. He runs West Coast Fire Training, a company that helps departments develop and deliver realistic live fire training in fixed facilities and acquired structures. He is a member of the NFPA 1400 committee and a technical panel member on the ULFSRI study on fire dynamics and firefighting tactics in multi-story residential buildings.

MIKE BERNARD
MASKING UP – Pg 87

Mike Bernard is a Captain and Training Officer with Arbor Vitae Fire and Rescue Department (WI). Bernard has been a student of the fire service for 15 years and is a contributor/instructor with Midwest Fire Tactics and Training. Mike is fortunate to share his passion for the job at local trainings and national conferences. When he isn't at the fire department he can usually be found enjoying the outdoors or loitering at a hockey rink watching his kids skate.

BRENT BROOKS
HIGH-RISE SEARCH AND RESCUE – Pg 169

Brent Brooks is an international speaker and hands-on instructor who teaches high-rise tactics and large-diameter hose movements. With 30 years of experience, he currently serves as an Acting District Chief with Toronto Fire Services and is assigned to Toronto's High-Rise Unit. Brent has developed the IMS, training, and R&D for High-Rise Operations and has extensive experience serving on numerous committees related to high-rise firefighting globally. He is well sought after for his presentations and hands-on training. Additionally, Brent contributes to developing safety guidelines for tall buildings

BRIAN BRUSH
COMMAND FOR THE RESCUE – Pg 139

Brian Brush began his career in 2002 at West Metro Fire Rescue in Lakewood, Colorado. Currently the training chief at Midwest City Fire Department. He holds a master's degree from Oklahoma State University, is a designated chief training officer through the CPSE, and is a graduate of the National Fire Academy EFO Program. An author for Fire Engineering since 2009, an instructor at FDIC since 2011, and FDIC keynote speaker in 2023, Brian is a Fire Engineering editorial advisor and serves on the FDIC advisory board.

AKA: Brian Crush.

AUTHOR GLOSSARY

AUTHOR GLOSSARY

MIKE CIAMPO
TOWER LADDERS FOR RECUE AND REMOVAL – Pg 245

Michael N. Ciampo is a retired Lieutenant of the Fire Department of New York (FDNY) where he spent 34 years. He first joined the fire service in 1979 as a fourth-generation volunteer firefighter in Wyckoff NJ. He attended John Jay College of Criminal Justice, NY and earned a BA degree in fire science. He is a nationally recognized instructor teaching truck company operations and tower ladder classes to departments across the country. He writes the monthly back page column ON FIRE for Fire Engineering magazine and is featured in their Truck Company Training Minutes videos. He is the author of the textbook Tower Ladders, Tactics, Tips & Tales available at mikeciampo.com

SCOTT CORRIGAN
THE FIRST IN INTERVIEW – Pg 131

Scott Corrigan is a Division Chief of Training in Gig Harbor, Washington. Previously he was an Engine Company Officer and Firefighter with 20 years' experience. He is also an instructor for First Due Training in Washington State providing training in auto extrication, fire behavior, engine company operations, small unit leadership, and tactical ventilation. He has delivered that training nationally, and internationally. He has written articles on fireground ventilation, leadership, and culture. He was instrumental in the development of fireground SOP's, and supportive training for his agency. He was a technical panel member for the UL FSRI PPV experiments.

CHAD DAILEY
LARGE AREA SEARCH – Pg 155

Chad Dailey is a Battalion Chief with the Kansas City, Missouri Fire Department and an owner of O'Byrne Fire ground Training. He joined KCFD in 1991. He spent 15 years on Missouri Task Force 1. He was deployed with the Task Force multiple times on incidents such as: Ground Zero in New York City on 9/11, Hurricane Katrina, and Hurricane Sandy. Chief Dailey has been married for 32 years and has two daughters and two amazing grandchildren who are the light of his life. In his free time, he enjoys fishing, exercising, traveling and the lake "life".

JIMMY DAVIS
HIGH-RISE SEARCH AND RESCUE – Pg 169

James "Jimmy" is a captain, and 29-year veteran of the Chicago (IL) Fire Department assigned to Engine Co. 43 serving Logan Square. He teaches regularly at the Chicago Fire Department R.J. Quinn Fire Academy on a variety of subjects, including the department's high-rise and standpipe programs. He instructs both nationally and internationally on many subjects relating to high-rise firefighting strategies and life-safety concerns.

AUTHOR GLOSSARY

AUTHOR GLOSSARY

MO DAVIS
PRIMARY SEARCH VS EMERGENCY RESCUE MODE – Pg 45

Maurice "Mo" Davis is a veteran with the City of Houston Fire Department, Currently assigned as District Chief. He has served as a Sr. Captain on a Heavy Rescue, one of three rescue trucks for the City of Houston. He also spent much of his career as a Sr. Captain on L-46 one of the busiest trucks in Houston. While he specializes in technical rescue, his passion is sharing leadership, strategy and tactics information with new and upcoming officers. He has served as an adjunct professor at San Jacinto Jr. College in the fire protection division. He currently holds a MS degree from the university of Houston.

MATT DONEY
WINDOW BASED RESCUE – Pg 191
WINDOW CONVERSIONS – Pg 205

Matt Doney is a 22-year veteran in the fire service. He currently works for the City of Porterville Ca as a Truck boss. Matt has always been passionate about Truck work and has been a member of the Brothers in Battle cadre since 2016 where he teaches Ladders, Basic and advanced Irons and all things search both nationally and internationally. "I believe fighting conditions and searching for life is gods work. I love Truck work and will pass it on to anyone who is willing to listen and put in the work to make it happen on the fire ground."

SHAWN DONOVAN
GROUND LADDERS FOR RESCUE – Pg 61

Shawn Donovan is a Captain on the Boston Fire Department. He has been a member of Boston Fire for 22 years. Shawn has slowly made his way through the ranks. Positions he has held include Firefighter and Lieutenant on Rescue 1, Lt of Ladder 4 and Engine 53, Fire academy instructor and Firefighter on Ladder 2. Shawn is also an instructor with Take the Door and has been a lecturer & HOT instructor multiple times at FDIC. He has two adult daughters and is enjoying the empty nest life

SEAN DUFFY
SEARCHABLE SPACE: THE SIZE UP – Pg 127
SPLIT SEARCH - Pg 231

Sean Duffy has been a firefighter for over 20 years, he is a Firefighter with the City of Ann Arbor Fire Department. Sean became involved in the training role and now leads one of the most recognized and sought after search programs in the country. Sean is a Co-founder of Build Your Culture, a published author, national speaker and has instructed all over the country and internationally. Sean was recently invited to serve as a technical panel member for the UL FSRI Multi-Story Residential Tactics Research Project. He believes strongly in providing quality fire training as well as maintaining a desire for continuous growth, both professionally and personally.

AUTHOR GLOSSARY

AUTHOR GLOSSARY

BOBBY ECKERT
SEARCHING OFF THE INITIAL ATTACK LINE – Pg 251

Bobby Eckert. Second Generation Fireman, from Collingswood NJ.
Fire Captain, Camden, NJ.
Owner/Operator Eckert Fire Tactics LLC.
Father to Bobby, Ryan, Kennedy, and Reese.

Let's Go Flyers!

JEFF ECKERT
THE DIRTY SIDE OF THE PRIMARY SEARCH – Pg 101

Jeff Eckert has been in the fire service for 26 years. The last 19 years he has served in the Philadelphia Fire Department. He has been a member of Special Operations for 14 years and has been assigned to Squad 72, for the last 12 years. He is co-owner of Philly Tin Helmet Training, A Training company that bases their classes on real world firefighting experiences. Jeff is a husband and a father to two children.

ROB FISHER
SEARCHING WITHOUT A HOSELINE – Pg 117

Rob Fisher is a suburban firefighter from the Pacific Northwest. He has been a student of the fire service for 36 years. Rob spent 20 years as a company officer primarily assigned to the department's only ladder company. In addition to Brothers in Battle, Chief Fisher instructs with County Fire Tactics, Roof Pervs, and the Highwaymen. He also co-founded Young Officers on Fire, an online mentoring group. Chief Fisher has been an active FOOLS member since 2002 and has been on the International's E-board since 2013. When not instructing or talking shop, Rob enjoys traveling with his wife of 31 years, Jodie. He enjoys cooking, wine, music, and spending time with his two granddaughters.

BASSEL IBRAHIM
TRAINING AGAINST THE CLOCK – Pg 65

Bassel Ibrahim is a 20 years student of the fire service and currently serves as the Company Officer of Tower 1 for the Orlando Fire Department. He has spent the majority of his time on Special Operations units. Bassel is the founder of the Firefighter Mayday Survey and is a certified Florida and Georgia Smoke Diver. He is a graduate of Georgia F.L.A.M.E.S. and the owner and lead instructor of Proactive Rescue Operations (PRO), a training company that specializes in Rapid Intervention Teams (RIT), Survival, Truck Operations, and Leadership. He is a successful author, lecturer, and podcast guest. And is highly sought after for his presentations.

AUTHOR GLOSSARY

AUTHOR GLOSSARY

JAMES JOHNSON
THE IMPACT OF ERA ON SEARCH – Pg 161

James Johnson is a firefighter with Vancouver Fire and Rescue Services and currently assigned to Squad 7 in Vancouver's downtown. James is a Building Codes and Standards Consultant for the International Association of Fire Fighters as well as a Master Instructor for the IAFF's Fire Ground Survival Program.

James is also a Technical Panel Member for ULFSRI's Fire Dynamics and Firefighting Tactics in Multi-Story Residential Structures research project.

TIM KLETT
SEARCHING IN HOARDER OCCUPANCIES – Pg 201

Tim is a 42 year Fire Service veteran serving the last 31 years with the FDNY until his retirement. Tim was a Lieutenant on Engine 88 in the Bronx for 19 years after spending many years on Engine 69, one of FDNY's busiest Engines. Tim is the lead HOT instructor for FDTN Academy's engine operations program and is the author of FDTN's "Stretching and Operating the first line" Tim has been lead instructor for FDIC and has taught engine operations at Firehouse Expo, the Orlando Fire Conference, Wichita HOT, and too many other conferences to list. Tim has written articles for Fire Engineering, FDTN, and Firehouse for many years. Tim was also a member of the group that totally re-wrote the FDNY's Engine Company Operations Manual.

CLARK LAMPING
HIGH-RISE SEARCH AND RESCUE – Pg 169

Clark Lamping is a 26-year veteran with the Clark County (NV) Fire Department where he has been a captain for 15 years. He has a bachelor's degree of Psychology from the University of Nevada, Las Vegas and a master's degree in Crisis and Emergency Management also from UNLV. He teaches highrise, midrise, and large commercial firefighting both nationally and internationally.

GARY LANE
SEARCH POSITIONS AND POISE – Pg 77
LIFE, FIRE, LAYOUT (AND CALLOUT!)– Pg 235

Gary Lane started in the fire service in 1997 as a volunteer. He has been a career fireman and paramedic in his hometown of Kent, Ohio since 2003. Gary is also an instructor with the Brothers In Battle training group.

AUTHOR GLOSSARY

AUTHOR GLOSSARY

DAVE LEBLANC
THE VIGILANT MINDSET – Pg 17

Dave is currently serving as the Fire Chief in Harwich Massachusetts. He began his fire service journey in 1986. He started his career in Harwich in 1993 where he rose through the ranks to become Fire Chief in 2020. Dave has shared his thoughts on firefighting and the fire service on Backstep Firefighter, FirefighterNation and Data not Drama. He has also taught at FDIC, Art of Firemanship and Training Days.

NICK LEDIN
COMMUNICATION CONSIDERATIONS – Pg 147

Nick Ledin is a captain with Eau Claire (WI) Fire Department, currently assigned to Truck 8. Ledin has been a student of the job for 17 years and is a contributor to 'Firefighter Rescue Survey', was the former president of the Northland FOOLS, and is a board member of FireNuggets. He's a technical committee member for NFPA 1700, was a former technical panel member for UL FSRI's PPA/PPV Study and is a co-host of the FireNuggets Podcast and GRABS Podcast. Nick is also lucky enough to be a small part of UL FSRI's Public Education Advisory Committee. He can also grow a pretty legit 'stache.

FRANK LEEB
SEARCH CULTURE – Pg 33

Frank Leeb is the managing director of the First Responder Center for Excellence (FRCE). He retired in June 2024 after more than 31 years in the FDNY. During his tenure with the FDNY, he held several senior staff positions, including the Chief of the Fire Academy, Chief of Training, and Chief of Safety. He has lectured nationally and internationally on motivation, preparedness, leadership, strategy, and tactics. He holds a bachelor's degree in fire service administration from SUNY Empire State and a master's degree in security studies from the Naval Postgraduate School, Center for Homeland Defense and Security (CHDS). He is also the author of the best-selling book "Cornerstones of Leadership - On And Off The fireground".

BRYAN LLOYD
SEARCH AND COMMON SENSE – Pg 37

Bryan Lloyd currently serves as the Chief of Training for the Tulsa Fire Department where he has worked for the last 24 years. Prior to starting his career in the fire service, he served in the U.S. Army for six years. He is the founder of the Oklahoma SmokeDiver Association and currently serves as the Chief Elder. Bryan is passionate about the fire service. He believes it is important to continue learning until the day you walk out of the station for the last time and to share what you know while continuing to learn from others regardless of rank or position.

AUTHOR GLOSSARY

AUTHOR GLOSSARY

MIKE LOMBARDO
WHY WE EXIST – Pg 13

Mike Lombardo is a 47-year veteran of the fire service and was the Commissioner/ Chief of Department of the Buffalo Fire Department from Jan 2006 to Jan 2010. Mike was a member of the development team for the New York State Firefighter Survival and Rapid Intervention programs. He is a two-time Firehouse Magazine Heroism Award recipient (1988 and 1994) and the recipient of the FDIC 1999 Tom Brennan Training Achievement Award. In 2015 Mike was a contributor for the NFFF documentary "Giving Courage: LODD Chiefs Speak". Mike currently operates with the Christiana Fire Department in Delaware and the East Amherst Fire Department in New York.

JUSTIN LORENZEN
RECORDING OUR RESCUES TO DRIVE OUR MISSION – Pg 49

Justin "J Lo" Lorenzen started in fire service in 2005 and he is currently a Lt./Paramedic on the south side of Oklahoma City at station 7/C, "Capitol Hill Creepers", where he drives Battalion 603. Justin served as a training officer OKCFD's training division. He's currently the president for OKC's local F.O.O.L.S chapter. He developed and implemented the OKCFD Grabs Survey that went live in 2020 allowing the Oklahoma City fire department to track their civilian rescues and contribute to Firefighter Rescue Survey. His passion is Search and all things training and most importantly leaving the job better than he found it!

DUSTIN MARTINEZ
KEEP SAYING YES TO VES – Pg 107

Dustin Martinez is a 15 year student, currently serving as a captain at Cobb County Fire Company 17. He is a byproduct of knowledgeable mentors meeting a happenstance of circumstances. He has instructed at local and national conferences and made more mistakes than his supervisors know about.

DAVE MELLEN
ORIENTED SEARCH – Pg 225

David Mellen is a 25-year student of the fire service and has been a paramedic for 19 years. David currently serves as a Captain with the Edwardsville Fire Department, a suburb of the Kansas City metro area, as well as battalion chief with Leavenworth County Fire District #2. He has sat on several review committees for fire training publications, presents at national, regional, and local fire conferences, and is the owner of Valor Fire Training. David lives in Shawnee, Kansas with his wife Katie, three sons, Aidan, Leo, and Jake, and their two dogs, Halligan and Quint.

AUTHOR GLOSSARY

AUTHOR GLOSSARY

JUSTIN MCWILLIAMS
THE WHY OF ENTERING WINDOWS – Pg 41
RULES OF ENGAGEMENT FOR SEARCH – Pg 113

Justin McWilliams started in the fire service in 2000. He's currently the Captain of Station 5 for Clackamas Fire. His department is unique in that they have multiple dry houses (No Engine/Pumps), which allows them the opportunity to gain experience of searching ahead of the hoseline often. He is a host of Grabs Podcast, an original member of Firefighter Rescue Survey (FFRS), a member of the ULFSRI Rescue & Search study, a cadre Member of Brothers in Battle, a father, husband and runs Search Culture socials page.

NICHOLAS PAPA
OPTIMIZING VENTILATION FOR SEARCH– Pg 173

Nicholas Papa is, first and foremost, a husband and father of two. He is a firefighter with 21 years of experience in the fire service. Currently a captain with New Britain (CT), where he has served for 17 years, and is presently assigned to the Training Division. He was the co-founder of the department's fire academy, and the lead instructor for the inaugural recruit class. He is a contributor to Fire Engineering and the author of the best-selling book, Coordinating Ventilation: Supporting Extinguishment & Survivability. Nick is a FDIC advisory board member and has been a classroom instructor since 2017. Nick is also the founder of the training and consulting organization, Fireside Training, LLC.

JON QUON
SECONDARY SEARCH – Pg 255

Jon got his start in the fire service as volunteer in his hometown in California. After many adventures he joined the Tacoma fire department where he is assigned as the Driver on Ladder 1. He has been blessed with the opportunity to share his love for the craft as a member of the Brothers in Battle cadre. He is an Indiana Smoke Diver. Jon is a proud father and husband.

ROB RAMIREZ
EFFECTIVE ARTIFICIAL STRESS – Pg 81

Rob Ramirez is a 23-year fire service veteran with degrees in Fire Science and Public Administration. Since 2005, he has been a Rescue Team Manager for F.E.M.A. U.S.A.R. Task Force in Miami, FL. where he has served on many high-profile national and international deployments. His education and personal experiences have provided him with a variety of unique instructional strategies that apply to today's Firefighters. Rob has held the ranks of Firefighter, Paramedic, Lieutenant, Training Officer. Captain. Rob is the proud father of two amazing boys and is currently assigned to a busy Truck company in Broward County, Florida.

AUTHOR GLOSSARY

AUTHOR GLOSSARY

JEFF ROTHMEIER
CRUCIBLE MOMENTS – Pg 21

Jeff Rothmeier has spent 18 years in the fire service; he currently serves with the Milwaukee Fire Department as a proud member of Rescue 1. During this time five years were as a company officer. Former member of the Minnesota Aviation Rescue Team. Former member of Minnesota Task Force 1 State Urban Search and Rescue (USAR) and structural collapse team.. Author of Mastering The Craft: Cultivating a philosophy for fire attack, Contributing Author for Fire Engineering and Fire Rescue., and a Proud, Decorated Combat Veteran!

RYAN SCELLECK
THE PSYCHOLOGY AND PHYSIOLOGY OF SEARCH – Pg 97

Ryan Scellick is a 24 year veteran of the fire service. He currently serves as senior shift Captain on the only ladder truck in a five station suburban department, following a three year period as the training officer. He has served on the department's land technical rescue team for six years. Ryan spent more than five years as commissioned SWAT Officer and tactical medic. Ryan is also the co-founder of Young Officers on Fire where he teaches leadership and training concepts nationally.

GRANT SCHWALBE
CIVILIAN DRAGS – Pg 181
ORIENTED VS. SPLIT SEARCH – Pg 217

Grant has been a firefighter since 1995. He moved to Florida in 2003 to work for Estero Fire Rescue (Lee County, FL) and spent 16 years as a Lieutenant on E43. In 2021 Grant was promoted to Division Chief of Operations and Training. Grant is an instructor who has taught at Conferences across the country. He hosts The Grabs and The Journeyman Firefighter Podcast. He is on the technical panel for the FSRI Fire Dynamics and Firefighting Tactics in Multi-Story Residential Structures. He's married to Katie and has 2 daughters (Lindsey and Peyton).

PAUL SHOEMAKER
LITHIUM ION & SEARCH REMOVAL – Pg 93

With 17 years of dedicated service in the fire industry, Paul is currently a Firefighter Paramedic based in Colorado. His commitment to enhancing firefighter safety extends beyond the frontline. He contributes to advisory panels with UL and collaborates with NFPA on a variety of projects ranging from electric vehicle extrication to addressing the challenges posed by lithium battery fires. As the founder of the Next Level Training Network, he is passionate about empowering firefighters nationwide with the knowledge and skills necessary to navigate the complexities of electric vehicles and lithium battery incidents.

AUTHOR GLOSSARY

AUTHOR GLOSSARY

BEN SHULTZ
THE IMPACT OF TIME – Pg 25

Ben Shultz is a lieutenant/ paramedic with the West Palm Beach Fire Department. He began his career in 2004 in southern California, relocated to Colorado, and eventually to Florida. He has worked in rural, suburban, and urban districts, and has split his time riding the truck, engine, heavy rescue, squad and medic. In addition to his normal shift duties, Ben has been a member of wildland, dive, swiftwater, and technical rescue teams, including FEMA CO-TF1. Ben has been fortunate enough to instruct with Irons & Ladders, When Things Go Bad, Vent Enter Search, and Fire By Trade.

ANDY STARNES
THERMAL IMAGING TIPS FOR SUCCESS – Pg 69

Andy Starnes is a lifelong student of the fire service since 1990. He is a fire service expert on the topics of thermal imaging, fire behavior, leadership, behavioral health, and faith-based devotions. He is passionate about Fire Behavior and assisting in instructing others on understanding the complex terminology of modern fire behavior on a regular basis. He serves as a board member on the non-profit organization 5-Alarm Task Force. Andy is a Master Craftsman with Project Kill the Flashover, and supports numerous first responder charities through Insight Training. Andy Starnes was inducted to the Hall of Legends, Legacies, and Leaders by the National FireHeritage Center at FDIC in 2023.

DAVID STONE
CONSIDERATIONS FOR VICTIM REMOVAL – Pg 187

David (DJ) Stone is a 2nd generation firefighter and a 23 year veteran of the fire service, currently serving as the Training Division Chief for South Walton Fire District. He has also worked for 4 other FD's (part-time) positions ranging from suburban to rural settings. David has instructed at various conferences including FDIC, HROC, The Fireground Commander Conference, Orlando FC, Metro Atlanta FC, Gone to Texas Fire Forum, as well as many other regional training events. DJ Stone has been an adjunct HOT instructor with many other teams specializing in suburban engine company operations.

SHANNON STONE
THE SUBURBAN ENGINE SEARCH – Pg 241

Shannon Stone is a second-generation 34-year veteran of the fire service who is working his way through the ranks. He holds multiple degrees and State certifications and has been a Florida State Paramedic since 1997. Shannon teaches at various events, including the Orlando Fire Conference, Metro-Atlanta Fire Conference, Portland Firemanship Conference, Lone Star Fire Conference, HROC, Command Officer Boot Camp, and FDIC. He specializes in teaching company officer leadership and engine company operations. He is currently with the Midway Fire District in Gulf Breeze, Florida, and was promoted to Fire Chief in August 2024.

AUTHOR GLOSSARY

AUTHOR GLOSSARY

SCOTT THOMPSON
THE SEARCH: A CHIEF'S PERSPECTIVE – Pg 29

Scott Thompson entered the fire service in 1982. Over the course of his career, he has worked in operations, training, and administration. Scott has been the fire chief in The Colony, Texas since 2009. He has been a hands-on, workshop, and classroom presenter at FDIC since 2002. Chief Thompson has had numerous articles published and is the author of The Functional Fire Company: Positioning Small Groups for Success and Survival. He owns and operates Functional Fire Company LLC and fireserviceleadership.com. His two sons are on the job and work in Arlington, Texas.
Reach him by emailing: scott@fireserviceleadership.com

ERIC WHEATON
CAN CONFIDENCE– Pg 211

Battalion Chief Eric Wheaton is a second-generation firefighter with the Winter Park (FL) Fire Department, following in the footsteps of his father. Eric has dedicated the majority of his career to the Truck Company and is a member of the WPFD Technical Rescue Team. Eric is the owner of the training company VentEnterSearch, LLC., where he presents on the national level, speaking and instructing at various conferences. He also serves as an advisory board member on the UL-FSRI Size-up and Search Rescue Study. Residing on eastern coast of Central Florida, Eric enjoys spending his free time outdoors with his wife and three children surfing, traveling, hiking, and boating. He has a strong passion for health and fitness.

DANE YAW
ASKING THE RIGHT QUESTIONS – Pg 135

Major Dane Yaw embarked on his public safety career over 30 years ago, beginning as a Police Paramedic before transitioning to the Oklahoma City Fire Department (OKCFD) in 1999. Over his career, Major Yaw has been instrumental in establishing and developing the Oklahoma Task Force One (OKTF-1) K-9 program. His extensive technical skills encompass Rescue Diving, Swift Water Rescue, Confined Space Rescue, Structural Collapse Rescue, Trench Rescue, and Rope Rescue. Currently, Major Yaw serves as the 606-A District ride-out officer with OKCFD, His career reflects a commitment to advancing the field and ensuring the highest standards of service and readiness in public safety.

AUTHOR GLOSSARY

THE ARTISTS

PAUL COMBS – COVER ARTIST

Paul an award-winning fine artist, illustrator, cartoonist, and firefighter/EMT. He is the editorial cartoonist for Fire Engineering Magazine, author of four books, and illustrator of the best-selling Sprinkles the Fire Dog series. He joined the fire service in 1995 and is now a retired Lieutenant for the City of Bryan Fire Department in Ohio. Paul is an FDIC presenter, keynoter, and Hands-On-Training instructor, and recipient of the 2018 George D. Post ISFSI Instructor of the Year Award. He is currently an instructor with On-Scene Training Associates Paul proudly serves as a USO entertainer, traveling the globe to bring comfort to active, wounded, and retired U.S. and NATO soldiers.

AMBER CHRISTIAN - ILLUSTRATIONS

Amber has been blessed to serve for the past 18 years as both a volunteer with the Town of Ixonia and now full time with Lake Country Fire Rescue. With a passion for drawing that started from the time she could hold a crayon; Amber embraces every opportunity to blend her artistic side with her love of the job!

CHRIS SMEAD - PHOTOGRAPHER

Chris is a professional photographer based in San Jose, California, with a primary focus on firefighting. With over 17 years of experience behind the lens, Chris has honed what was once a hobby into a career. By photographing firefighters, Chris captures the bravery and heroism of the firefighters, offering a powerful glimpse into their world of selflessness and dedication. With this niche focus and opportunity, Chris hopes to capture images that will last a lifetime

THE ARTISTS

JILLIAN STEWART - PHOTOGRAPHS

Jillian is an Artist and Photographer based out of New Cumberland, PA. who got into fireground photography totally by accident. Jillian was helping her local department with putting together some recruitment material and she wanted to include fire photos in her marketing pieces. She took her first fire scene photos in 2017 and has been hooked ever since. Since then, she has been published in multiple books as well as being a featured photographer at the Pennsylvania National Fire Museum.

The effort Jillian devotes to her photography comes from her wanting to document not only the day-to-day life of firefighters, but also the random acts of kindness. In addition to photographing local fire scenes, Jillian also travels around the country documenting training conferences.

SAMANTHA WORTHY – PHOTOGRAPHER

Upstate South Carolina fire wife and the creative mind behind Camera Worthy Photography. Her journey into photography started in her childhood, inspired by her dad. Her deep appreciation for capturing the fire service came after she met her husband, Jason, a second-generation career firefighter. In 2019, she began traveling to document training events and conferences, which quickly turned into a passion. Over the past year, She has collaborated with local departments to capture their daily routines and calls, while also assisting with recruitment, marketing, and social media content. She loves being able to share in her husband's passion for the job and everything firefighting entails. Through her photography career, she has encountered some of the most humble, passionate, and dedicated individuals in the fire service, and it is truly a privilege to work alongside them.

AUTHOR SCRAP EPISODES

The following authors have appeared on the Scrap. This list will help you find their episodes so you can hear more of what they have to say!

Ian Bennett	242
Brent Brooks	220
Brian ~~B~~ Crush	23, 217, Mega Scrap Charlie
Paul Combs	140
Chad Dailey	127
Mo Davis	114
Shawn Donovan	76
Sean Duffy	58, 168 & Mega Scrap Echo
Bobby Eckert	125, 165 & 222
Jeff Eckert	194
Rob Fisher	131, 226, Host 260, Mega Scrap Delta
Bassel Ibrahim	98, 214 & Mega Scrap Bravo
James Johnson	33, & 207
Tim Klett	172
Clark Lamping	199
Gary Lane	51
Dave LeBlanc	53 & 143
Nick Ledin	40 & 138 & Mega Scrap Echo
Frank Leeb	183, 213, & Mega Scrap Foxtrot
Bryan Lloyd	256
Mike Lombardo	202 & 271
Justin Lorenzen	44
Dustin Martinez	107
David Mellen	84
Justin McWilliams	164 & 268
Nicholas Papa	109
Rob Ramirez	96, 108, & 184 & Mega Scrap Bravo
David Rhodes	132 & 234
Jeff Rothmeier	211
Ryan Scellick	249
Grant Schwalbe	177
Paul Shoemaker	215
Ben Shultz	57
Andy Starnes	59
David Stone	207
Shannon Stone	82, 203
Scott Thompson	27, 64, 118, 178, 244, Host 261 & Mega Scrap Alpha & Echo
Eric Wheaton	90
Dane Yaw	108 & 236

HISTORY

HOW THE BOOK OF SEARCH CAME TO BE

> "It is amazing what you can accomplish if you do not care who gets the credit"
> ~ Harry S. Truman

I believe the above quote is the reason why this book you are reading exists. There are a lot of selfless firefighters who are dedicated to improving this profession. This book was an amazing journey for me that lasted nearly eighteen months. What started as a small conversation with a fellow firefighter, Zach Golembeski out of Kansas, over a phone call (which I don't even know if he remembers) turned into a long and laborious journey to make this book a reality.

I've always considered myself "the luckiest guy in the fire service," and a major reason for that is the incredible network of firefighters I've built through the Scrap podcast. I knew my network was impactful, which fed the idea of making this book happen.

The book needed a bigger purpose— one that everyone could get behind and support. That's where Firefighter Rescue Survey comes in. I believe it's one of the most significant initiatives in firefighting over the past fifty years. What if we could donate all the book's profits to Firefighter Rescue Survey? Now that is an idea everyone could rally behind!

After determining where the profits would go, I reached out to Nick Ledin, my contact for Firefighter Rescue Survey, to pitch my idea. Here's the crazy part: I didn't realize that the team at FFRS and Fire Nuggets had started a similar project two years prior. Their idea was to create a "Search Manual," but they had lost momentum along the way.

I told Nick, "I am a gas pedal. It's what I do." I was off and running with the Book of Search.

Before my phone call with Nick, I had already collected several articles. Afterward, I began working with their original document, which contained 14,000 words across various rough drafts of

HISTORY

articles. While their project had stalled out, they had brainstormed an amazing list of topics. By combining that list with the network of the Scrap… give it eighteen months, throw in a lot of sweat, texts, emails and phone calls… and boom— over 110,000 words across forty-six articles from forty-three subject matter experts.

The result is the book you're reading now.

I would love to thank each author individually, but that would require a whole additional section of the book! Still, I want to share a few thoughts. I know I drove some of them crazy with my non-stop phone calls, texts and deadlines. I'd like to thank them for their patience and hard work as we pushed this book across the finish line. A special shoutout to those authors who promised an article and delivered it on time without needing a single reminder— after this process, I see how truly rare that is.

I want to give a special thanks to Amber, Chris, Jillian, Paul and Samantha. Your art is what made the book come alive. I am truly jealous of each of your creative talents!

Thank you to everyone out there who is fighting the fight to improve the fire service. I hope you find value in this book and that it inspires you in your journey.

Stay strong in the never ending fight against complacency!

Godspeed,
Corley Moore

Made in the USA
Columbia, SC
27 April 2025

2b3d3ebf-c78c-41cc-8cbd-09b05d2b69dcR02